Bioeconomical Solutions and Investments in Sustainable City Development

José G. Vargas-Hernández
University of Guadalajara, Mexico

Justyna Anna Zdunek-Wielgołaska
Warsaw University of Technology, Poland

A volume in the Practice, Progress, and Proficiency in Sustainability (PPPS) Book Series

IGI Global
DISSEMINATOR OF KNOWLEDGE

Published in the United States of America by
IGI Global
Engineering Science Reference (an imprint of IGI Global)
701 E. Chocolate Avenue
Hershey PA, USA 17033
Tel: 717-533-8845
Fax: 717-533-8661
E-mail: cust@igi-global.com
Web site: http://www.igi-global.com

Library of Congress Cataloging-in-Publication Data

Names: Vargas Hernandez, Jose G., editor.
Title: Bioeconomical solutions and investments in sustainable city development / Jose G. Vargas-Hernandez and Justyna Anna Zdunek-Wielgolaska, editors.
Description: Hershey, PA : Engineering Science Reference, [2019]
Identifiers: LCCN 2018041835| ISBN 9781522579588 (hardcover) | ISBN 9781522579595 (ebook)
Subjects: LCSH: Urban ecology (Biology) | City planning--Environmental aspects. | Sustainable urban development.
Classification: LCC QH541.5.C6 B54 2019 | DDC 577.5/6--dc23 LC record available at https://lccn.loc.gov/2018041835

This book is published in the IGI Global book series Practice, Progress, and Proficiency in Sustainability (PPPS) (ISSN: 2330-3271; eISSN: 2330-328X)

British Cataloguing in Publication Data
A Cataloguing in Publication record for this book is available from the British Library.

Practice, Progress, and Proficiency in Sustainability (PPPS) Book Series

ISSN:2330-3271
EISSN:2330-328X

MISSION

In a world where traditional business practices are reconsidered and economic activity is performed in a global context, new areas of economic developments are recognized as the key enablers of wealth and income production. This knowledge of information technologies provides infrastructures, systems, and services towards sustainable development.

The **Practices, Progress, and Proficiency in Sustainability (PPPS) Book Series** focuses on the local and global challenges, business opportunities, and societal needs surrounding international collaboration and sustainable development of technology. This series brings together academics, researchers, entrepreneurs, policy makers and government officers aiming to contribute to the progress and proficiency in sustainability.

COVERAGE

- Global Business
- ICT and knowledge for development
- Innovation Networks
- Sustainable Development
- Technological learning
- Green Technology
- Intellectual Capital
- Strategic Management of IT
- E-Development
- Eco-Innovation

The Practice, Progress, and Proficiency in Sustainability (PPPS) Book Series (ISSN 2330-3271) is published by IGI Global, 701 E. Chocolate Avenue, Hershey, PA 17033-1240, USA, www.igi-global.com. This series is composed of titles available for purchase individually; each title is edited to be contextually exclusive from any other title within the series. For pricing and ordering information please visit http://www.igi-global.com/book-series/practice-progress-proficiency-sustainability/73810. Postmaster: Send all address changes to above address.

Titles in this Series

For a list of additional titles in this series, please visit:
https://www.igi-global.com/book-series/practice-progress-proficiency-sustainability/73810

Optimizing the Use of Farm Waste and Non-Farm Waste to Increase Productivity...
Leighton Naraine (Clarence Fitzroy Bryant College, St. Kitts-Nevis)
Engineering Science Reference • ©2019 • 223pp • H/C (ISBN: 9781522579342) • US $155.00

Handbook of Research on International Collaboration, Economic Development, and ...
Vasilii Erokhin (Harbin Engineering University, China) Tianming Gao (Harbin Engineering University, China) and Xiuhua Zhang (Harbin Engineering University, China)
Business Science Reference • ©2019 • 703pp • H/C (ISBN: 9781522569541) • US $235.00

Handbook of Research on Rural Sociology and Community Mobilization for Sustainable ...
Qaiser Rafique Yasser (Centre for Rural Economy, Pakistan)
Information Science Reference • ©2019 • 393pp • H/C (ISBN: 9781522571582) • US $245.00

Impacts of Violent Conflicts on Resource Control and Sustainability
Esther Akumbo Nyam (Plateau State University, Nigeria) and Asiru Hameed Tunde (Umaru Musa Yar'adua University, Nigeria)
Engineering Science Reference • ©2019 • 218pp • H/C (ISBN: 9781522559870) • US $185.00

Practice and Progress in Social Design and Sustainability
Kin Wai Michael Siu (The Hong Kong Polytechnic University, Hong Kong) and Yin Lin Wong (The Hong Kong Polytechnic University, Hong Kong)
Information Science Reference • ©2019 • 330pp • H/C (ISBN: 9781522541837) • US $195.00

Handbook of Research on Methods and Tools for Assessing Cultural Landscape Adaptation
Isabel de Sousa Rosa (University of Lisbon, Portugal) Joana Corte Lopes (University of Lisbon, Portugal) Ricardo Ribeiro (University of Lisbon, Portugal) and Ana Mendes (University of Lisbon, Portugal)
Engineering Science Reference • ©2018 • 524pp • H/C (ISBN: 9781522541868) • US $265.00

For an entire list of titles in this series, please visit:
https://www.igi-global.com/book-series/practice-progress-proficiency-sustainability/73810

IGI Global
DISSEMINATOR OF KNOWLEDGE

701 East Chocolate Avenue, Hershey, PA 17033, USA
Tel: 717-533-8845 x100 • Fax: 717-533-8661
E-Mail: cust@igi-global.com • www.igi-global.com

Editorial Advisory Board

Table of Contents

Preface xv

Section 1
Conceptual and Theoretical Bio-Economy Framework or Reference

Chapter 1
Bio-Economy: Visions, Strategies, and Policies 1
Baseem Khan, Hawassa University, Ethiopia

Chapter 2
Strategies of Sustainable Bioeconomy in the Industry 4.0 Framework for Inclusive and Social Prosperity 21
Erhan İşcan, Cukurova University, Turkey

Chapter 3
Urban Farming in Sustainable City Development 43
Magdalena Grochulska-Salak, Warsaw University of Technology, Poland

Chapter 4
Parametric Evaluation of Beam Deflection on Piezoelectric Material Using Implicit and Explicit Method Simulations: A Study in Energy Engineering 65
Rakesh Nath, Girijananda Chowdhury Institute of Management and Technology, India

Section 2
Issues Related to Investment in Different Industry Sectors

Chapter 5
Entrepreneurial and Institutional Analysis of Biodiesel Companies in Mexico 89
José G. Vargas-Hernández, University of Guadalajara, Mexico
Juan José Esparza López, University of Guadalajara, Mexico
Justyna Anna Zdunek-Wielgołaska, Warsaw University of Technology, Poland

Section 3
Specific Cases on Sustainable City Development

Chapter 6
Ecological Centre of Warsaw as a Development Path 117
Krystyna Gruszecka, Warsaw University of Technology, Poland

Chapter 7
Green Spaces of the Metropolitan Area of Guadalajara 151
Roberto Jiménez, Universidad de Guadalajara, Mexico
Paula Lourdes Guerrero Rodríguez, Universidad De Guadalajara, Mexico
Rogelio Rivera Fernández, Universidad De Guadalajara, Mexico

Chapter 8
Application of UAV Technology to Planning Study on Chinese Villages in Guanzhong 180
Yuan Zhong Cai, Xi'an University of Architecture and Technology, China
Feng Wu, Xi'an University of Architecture and Technology, China
Jing Li, Xi'an Polytechnic University, China
Jin Wang, Xi'an University of Architecture and Technology, China
Mei Huang, Xi'an University of Architecture and Technology, China

Chapter 9
Developing a Sustainable Eco-City in Pre-Olympic Tokyo: Potential of New Methods and Their Limits in an Urban Era 196
Mariko Ikeda, University of Tsukuba, Japan

Chapter 10
Forest-River-Ocean Nexus-Based Education for Community Development: Aiming at Resilient Sustainable Society 224
Shimon Mizutani, Tokyo University of Marine Science and Technology, Japan
Kai Liao, Tokyo University of Marine Science and Technology, Japan
Tsuyoshi Goto Sasaki, Tokyo University of Marine Science and Technology, Japan

Glossary 249

Compilation of References 254

Related References 277

About the Contributors 302

Index 305

Detailed Table of Contents

Preface xv

Section 1
Conceptual and Theoretical Bio-Economy Framework or Reference

This section presents four chapters addressing some relevant issues focusing on concepts and theories of the bio-economy paradigm. The four chapters conceptually explore some different trajectories and contestations of this paradigm, arguing more broadly that this is becoming a significant driver for re-structuring corporate and competitive strategies and policies between firms, markets, and urban and rural spatial development between and within places.

Chapter 1
Bio-Economy: Visions, Strategies, and Policies 1
Baseem Khan, Hawassa University, Ethiopia

As the world's economy is transforming from conventional to sustainable mode, bio-economy will play an important role in this transition. The development of sustainable bio-economy is the great challenge in different parts of the world. For this purpose, adequate visions, strategies, and policies are required. Therefore, this chapter deals with the adequate visions, key strategies, and important policies for the development of sustainable bio-economy. Some case studies also presented where these visions, policies, and strategies are already implemented.

Chapter 2
Strategies of Sustainable Bioeconomy in the Industry 4.0 Framework for Inclusive and Social Prosperity 21
Erhan İşcan, Cukurova University, Turkey

The last two decades have witnessed unpredictable transformations in knowledge-related areas. The main reason for this change is the fourth industrial revolution, a knowledge revolution affecting fields like the bioeconomy. The third industrial revolution, which induced the use of fossil-based energy sources, created a major global problem. Likewise, the third industrial revolution introduced the problem of the excess usage of food, animals, water, and other resources. Industry 4.0 offers an efficient solution to excessive tendencies. This chapter aims to analyze changes and offer strategies in the bioeconomy framework within Industry 4.0.

Chapter 3
Urban Farming in Sustainable City Development 43
Magdalena Grochulska-Salak, Warsaw University of Technology, Poland

Urban farming is defined as building development for the production of plants for the needs of the local community. The presented issues concern the shaping of urban farms for the preservation of the environmental balance of urban spaces and connections concerning the coexistence of architecture and greenery in the city. This chapter indicates the possibility of shaping synergistic spatial systems by integrating urban space and buildings with an innovative production function—a municipal farm—that complements the functional structure of the city in connection with the shaping of public spaces and the greenery system. The pro-environmental architecture connected with technologies enabling the production of plants in buildings enables the integration of urban space, complementing the functional and spatial structure of the city. The implementation of new technologies enables the production of plants in hydroponic and aeroponic farm buildings. The urban farm is an element in planning the city's sustainable development.

Chapter 4
Parametric Evaluation of Beam Deflection on Piezoelectric Material Using Implicit and Explicit Method Simulations: A Study in Energy Engineering 65
Rakesh Nath, Girijananda Chowdhury Institute of Management and Technology, India

Energy harnessing methods from rainwater using various smart materials have come into research, thus widening the scope of using these materials for the implementation in modern use. The piezoelectric materials present a brief idea of voltage generation whenever the material is deflected. The purpose of the study is to visualize an interconnection in parametric analysis of piezoelectric effect based energy harvester using two different commercially available piezoelectric materials, PZT-4A and PZT-5H, through series of implicit and explicit method simulations of FEM on COMSOL and ANSYS. The dynamic loads of different rainwater droplets sizes are investigated analytically. To calculate the variation of different methods in terms of deflection and voltage output, the implicit and multi-body explicit dynamic simulations are implemented separately.

Section 2
Issues Related to Investment in Different Industry Sectors

There is increasing consensus about nature providing a wide range of benefits to people and the importance of incorporating these ecosystem services into resource investment and management decisions of different industry sectors centered on sustainable and ecological development. This section explores the developments, achievements, and possibilities of specific cases related to the bioenergy sector (biodiesel companies in Mexico) and the bio-education for community development.

Chapter 5
Entrepreneurial and Institutional Analysis of Biodiesel Companies in Mexico 89
José G. Vargas-Hernández, University of Guadalajara, Mexico
Juan José Esparza López, University of Guadalajara, Mexico
Justyna Anna Zdunek-Wielgołaska, Warsaw University of Technology, Poland

The objective of the chapter is to analyze the role of the institutions in the biodiesel industry, in order to know if there is a relationship with the quality and maturity of the same with the ventures. Starting from a literary review, the framework of the current situation is identified, covering aspects related to formal institutions, laws, rules, regulatory bodies, and the theory that supports the relationship between institutions and entrepreneurship. The chapter concludes that the institutions in Mexico have increased their maturity and incentive to increase the number of producers and distributors of biodiesel, thus taking advantage of the growing market.

Section 3
Specific Cases on Sustainable City Development

Bio economy has a high importance for generating environmental, ecological, and socio-economic impacts, especially in populated urban areas and peripheries. This last section presents four chapters related to the topic of sustainable city planning and development. Each chapter explores and analyzes a specific city: Warsaw, Guadalajara, Chinese Villages in Guanzhong, and Tokyo.

Chapter 6
Ecological Centre of Warsaw as a Development Path 117
Krystyna Gruszecka, Warsaw University of Technology, Poland

The purpose of this chapter is to present a new development path towards greening the city center of Warsaw in the context of the increasing density of building development in recent years. After the process of urban sprawl, there has been a tendency to concentrate buildings, especially in the city center. Currently, a new idea and the need to improve the climate in the city is being born. The aim is to convert dispersed green areas into a continuous network of corridors and green spaces, comprising the city center on both sides of the Vistula River. The future structure of green areas in the central part of Warsaw will be built as an element of bioeconomy. According to this new pattern of urban greenery, larger green enclaves will be connected by corridors created out of necessity along densely built-up streets. Such elements as green walls, green roofs with decorative greenery and food crops, pocket greenery, as well as urban farms (e.g., algae energy generation) will complement buildings, foster healthy environment, and create the opportunity to enjoy pastimes.

Chapter 7
Green Spaces of the Metropolitan Area of Guadalajara 151
Roberto Jiménez, Universidad de Guadalajara, Mexico
Paula Lourdes Guerrero Rodríguez, Universidad De Guadalajara, Mexico
Rogelio Rivera Fernández, Universidad De Guadalajara, Mexico

The analysis of some systems of green areas and public parks of the metropolitan area of Guadalajara, other cities of our country Mexico and Latin America, shows common problems such as the deficit of urban green spaces, insecurity, unemployment, and uncertainty with a social exclusion in these areas of stress. Likewise, the lack of economic value of the services provided by such natural systems as recreation is added. Together they are important factors in the allocation of territories destined to this use with respect to others that generate Urban speculation. Therefore, it is proposed to develop a typology of green areas appropriate to the needs of the metropolitan region. It will facilitate the production of inventories that estimate indicators of territorial cohesion, governance, economic profitability, social, environmental quality and innovation, as well as incorporating new technologies that improve geographic information systems and internet media that support management.

Chapter 8
Application of UAV Technology to Planning Study on Chinese Villages in Guanzhong 180
Yuan Zhong Cai, Xi'an University of Architecture and Technology, China
Feng Wu, Xi'an University of Architecture and Technology, China
Jing Li, Xi'an Polytechnic University, China
Jin Wang, Xi'an University of Architecture and Technology, China
Mei Huang, Xi'an University of Architecture and Technology, China

Driven by the state strategy of rural revitalization, Chinese rural areas receive unprecedented opportunities for development. However, China's Guanzhong region faces numerous problems in its rural planning research, such as 1) lack of terrain maps of most villages, 2) satellite maps collected from open platforms are inaccurate and fail to support a more detailed spatial analysis, 3) data and information are 2-dimensional, 4) data collection is inefficient. And, most villages consist of several village groups that are usually 400~500 m apart. Areas of Guanzhong are located on the plain, with low architectural height and an excellent environment of net clearance. In addition, there are no large-scale factors, mineral areas, and industrial facilities, which means low interference from the magnetic field. Compared with urban regions, such rural areas have a better work environment for UAV and better conditions of collecting needed data.

Chapter 9
Developing a Sustainable Eco-City in Pre-Olympic Tokyo: Potential of New Methods and Their Limits in an Urban Era 196
Mariko Ikeda, University of Tsukuba, Japan

This chapter intends to elucidate the emergence of sustainable urban development in Tokyo in light of the upcoming 2020 Olympics by exploring various administrative and commercial practices, such as urban development plan with rooftop and wall greening or river projects in the inner city. The research methods involved a review of the empirical literature, an analysis of existing statistical data, and a detailed examination of case-specific data collected in a field survey. This chapter concludes that since Tokyo Metropolitan Government encourages urban greening projects as a solution of urban heat island from 2000s, utilization of "green spaces" in the landscape design of commercial and office facilities is gaining attention. This chapter concludes that various practices for sustainable urban development in Tokyo, which faces a restructuring process in light of the 2020 Summer Olympics, exist and that some of these could be further developed by the private sector.

Chapter 10
Forest-River-Ocean Nexus-Based Education for Community Development: Aiming at Resilient Sustainable Society...224
Shimon Mizutani, Tokyo University of Marine Science and Technology, Japan
Kai Liao, Tokyo University of Marine Science and Technology, Japan
Tsuyoshi Goto Sasaki, Tokyo University of Marine Science and Technology, Japan

Bioeconomic research aims at developing a more resource-efficient and sustainable society that uses renewable biological resources to produce food, materials, and energy. Economic supremacy causes many problems, such as global warming, depletion of fossil fuels and natural resources, and loss of biodiversity. In order to build a more sustainable society with resource efficiency, it is necessary to discuss the institutional framework, which includes environmental assessment, environmental monitoring, biological resource management, human resources management, and education. This chapter examined the effectiveness of forest-river-ocean nexus-based education for community development (FRONE) in encouraging the sustainable use of biological resources. Combined with the adaptive cycle, FRONE is considered to have the potential to promote the sustainable use of biological resources. In the future, further bioeconomic research from the point of view of the education system will be needed.

Glossary 249

Compilation of References 254

Related References 277

About the Contributors 302

Index 305

Preface

INTRODUCTION: AN OVERVIEW OF THE SUBJECT MATTER

Bioeconomical Solutions and Investments in Sustainable City Development is the title of this book that we, José and Justyna as editors, offer to the academic and scientific community. It is inferred from the mere tittle that the main subject matter is sustainable city development. From this subject, two main tools are analyzed as variables that are directly related to sustainable city development and have an impact on results, the bio economical solutions and investments. Sustainable urban planning and development is a methodological framework as a toolkit to transform a vision into reality using urban spaces as resources to engage stakeholders for achieving sustainable city development. A brief analysis of the three variables related and involved as the main axes of the book are briefly analyzed below.

Sustainable city development strategies are supported by urban planning practices based on urban governance involving stakeholders under the assumption that urban growth is interdependent with economic growth and can be reconciled in participatory and sustainable planning. As part of the integral urban spatial plan, some sustainable city development planning objectives should be attached to contemporary urban development trends. A sustainable city development can be achieved through an urban planning process (UN 2009) that responds to the requirements of sustainable urban development is needed for the sake of sustainable cities.

Regarding bio economical solutions to sustainable city development is sustained by the recent developments of bio economics. The term bio-economy is related to the role of sustainable biomass. Bio-economy is also named bio-based economy, although bio-economy refers to food and feed chains and bio-based economy refers to the bio products of non-food goods. However, bio economy is more useful term to mean both the use of food and feed, and the bio production. Also, aquatic biomass expands production for the bio-economy. Bio-economy activities are connected with sustainability inherently fossil-free shifting away from oil-based production and using more biomass and bio-based materials where no waste is produced.

Bio-economics is the set of economic activities that obtain products and services, generating economic value, using, as fundamental elements, resources of biological origin, in an efficient and sustainable way. In bio-economics, viability is a concept that implies time, context, and the nature of economic value. Bio-economy is characterized by the creation and efficient use of natural and biological resources, raw materials and capabilities in sustainable infrastructures aimed for the bio production of goods, bio services, bio energy, bio health, etc. to achieve sustainable lifestyles, wealth and economic growth.

One of the objectives of bio-economy is to integrate economics and environmental sciences associated with the use and conversion of biomass to bio production. The premise of bio economy is based on achieving a balance between economic activities and the use and management of biological natural resources. The Bio-economy addresses some major environmental, economic and social changes for sustainable production and transformation of biomass material for better living and working. Integrated bio-economy provides better living conditions for a sustainable human development, better food, raw material for bioenergy and bio products while halving the environmental impact.

Bio economics begun with rural-urban food chains and food banks promoting green diets. Integrated bio-economy secures transition to sustainable future by creating biomass as a renewable raw material for bio production and food security provided by agriculture. Sustainable bio-economy supports a transition from fossil economy to exploitation of renewable natural resources and biomasses produced by forests, waters and fields. In some countries such as Finland, bio-economy relies in forests based on sustainable exploitation. Transition towards a bio economy should be supported by policy instruments to achieve a sustainable economy and political stability. Legal and regulatory frameworks may facilitate organizational transitions towards a more bio economy-oriented activity in the market place and become positioned in a market share.

Bio-economy responds to current developments and challenges in global economics, social and environmental issues. Bio economic and environmental analyses are strongly correlated. Bio-economy has a relevant contribution to economic growth with the production of multiple public and social goods such as supplies in food, energy, coastal, rural production, and conservation of natural and biodiversity environments.

There are some value opportunities in bio-economy from producing high value products from using waste resources and feedstock. Bio economy sectors in economic activities have increased in wood-based industry, consumer goods industry, bio-based plastics, energy, etc. Bio-economy is not restricted to waste feedstock which also takes on process and harvest residues not produced as bio products but as by-products, co-products and biogenic components of industrial and consumer biodegradable bio-waste.

Bio-based products are products that are wholly or partly derived from materials of biological resources origin, excluding materials embedded in geological formations and/or fossilized. Bio-economy implies the capacity to develop a long term vision for the future use of bioenergy and bio resources, the production of renewable biological resources or biomass to provide more food security and a better life conditions to future generations.

Thus, spatial bio economics can support a modeling approach of efficient decision making mechanisms. The managerial control of spatial bio-economic models is related with restoration, development, harvest of habitats in any location and the dispersal and diffusion of species across the space.

The social importance of bio-economics seeks to facilitate access to basic services. Economic adaptability of bio economy is understood as the ability of social actors to produce and maintain maximum value over productive capacity in the territory, by strengthening links between sectors, combining assets to enhance the specific character of local products and services.

The other variable, investment and its implication in sustainable city development is sustained on the assumption that sustainable urban planning can foster economic development by promoting urban economies of scale and agglomeration, adaptation to climate change impacts, reducing use of energy, etc. Sustainable city development requires a realistic financial scenario with the sources of investment and mechanisms of cost-recovery to ensure social affordability and financial sustainability. Sustainable city planning and development is linked to budgeting in terms of investments in infrastructure and services to balance economic growth, social equity and environmental challenges.

Some challenges for institutional investors are represented by the development of green bond markets, investments in low-carbon and climate-resilient infrastructure, greening and achieving inclusion and equity in the public and private financial institutions and banking services, etc.

Sustainable urban planning and development processes must link economic development concerns to investment planning in infrastructure development. Sustainable urban planning and development infrastructure investment decisions can improve the living conditions of population over the long term. Sustainable city planning and development supports local economic development and coordinates the urban spatial locations and the efficient distribution of economic, social and environmental activities achieving value capture from public investments. The participatory planning processes have influence on environmental investments to respond to local environmental and sustainable city development.

It is also important to consider the investments in human capital. Bio economy-based companies have to train their working forces with specific technical issues and challenges related to industry sector besides training with specific managerial skills, change management and other techniques to be used to enhance the sustainable city development. Pioneering business organizations facing these challenges are training and developing the next generations with creative and co-operative knowledge and entrepreneurial skills to use it as innovation drivers and venture builders.

A DESCRIPTION OF WHERE THE TOPIC FITS IN THE WORLD TODAY

Spatial development is a discipline aimed at protection of specific values and rational development by stimulating economic processes. It is a domain that combines different groups of interests operating on one area. Its interdisciplinary nature arises from its definition; therefore, the solutions developed in spatial development have the impact on many users. Especially difficult and complex matter is spatial development and policy for large agglomerations, regarding modern challenges to minimize the negative impact of urban development on the environment. One of the ways to really impact quality changes in this respect is to strengthen the potential of bio economical solutions and investments.

Quality changes in city perception and its role in civilization progress occur in front of us. The issue of self-sufficient residential units is being increasingly discussed and in higher scale, such as the whole urban agglomeration. One of the ways is to introduce innovative solutions that can result in more radical and quality changes in the future. The political will, social awareness, financial resources and physical location will be necessary to implement these solutions.

This work is to collect different aspects related to development of sustainable city models based on investment in eco-oriented solutions that is literally, by protecting and making publicly available green areas and by innovative investments with the use of bio economical solutions. These solutions may be essential for both large agglomerations and shrinking cities. Bio economy and investments to develop green urban areas can be successfully used in both situations – as a way to reduce the demand for supply of external resources and step reduction of operating costs in the future. The main objectives of the book are to expand knowledge on possible implementation of innovative solutions in cities and nearest environment and to highlight a role of green areas in sustainable development of urban setting.

A DESCRIPTION OF THE TARGET AUDIENCE

Given the interdisciplinary nature of issues in question, they are directed to urban planners, architects, city authorities, experts, officers, business representatives, economists, politicians and academic centers and scientists. It can be used both as a source of practical information and to start common initiatives by different centers. The book will provide insights and support executives concerned with the quality of built sustainable environments.

THE CHALLENGES

In the history of the planet earth, the Homo sapiens has been the great predator of nature. Nowadays, the planet earth is confronting several grand and complex challenges in rural and urban development such as climate change, energy security, food security, resource depletion, etc., difficult to solve at the point to put at the brink the mere existence of high percentage of living natural species, including human beings.

The benefits arising from the economic growth in urban development have been accompanied by serious environmental degradation, social inequalities and production challenges. Some major environmental challenges to overcome are the increase of human population, over consumption, depletion and decoupling of natural resources, unsustainable economic development and growth, impacts on environment and ecology, global warming. Other challenges are related design of a wide range of feedstock characteristics, availability and quality, processing paths, product portfolios and trade-offs with energy consumption and production, distribution and market prices (Tsakalova et al., 2015).

The bio economy and the related biotechnology industry face outstanding and large technical challenges for sustainable city development production of high-value biomass or bulk green products. Renewable biotechnology is lagging behind when compared to developments of green, renewable chemistry in production of commodity chemicals. The alliance and convergence of green chemistry with industrial biotechnology is already solving some challenges in sustainable city development.

Sustainable city development challenges of scale for green chemicals are more easily surmounted than for other commodity chemicals. For example, biotechnology solutions to aromatics are particularly challenging. The aromatics challenge reinforces solutions for biotechnology that are lagging behind those for chemicals. Companies of renewable chemistry should have solutions to the aromatic challenge through process of non-food biomass such as corn stover, sugar cane bagasse, wood, sawdust, etc., are gasified and converted into hydrocarbons identical to the petroleum-derived products.

There are also some technical challenges to sustainable city development in perfecting processes when using waste materials as feedstock. Different degrees of uncertainty and complexity make it challenging for diverse industry sectors, business investors, authorities and officials from local governments, decision makers, etc., to identify the most promising bio economy options, including their technological and economic risks.

Different industry sectors that have been conducting research and innovation in bio economy development have been operating in isolation. Industry is struggling to produce bio-based products and chemicals at a scale influencing the market. Bio-based production challenges regulations across boundaries. Fossil-based feedstock are already unsustainable means of the production system that requires a gradual replacement with bio-based ones. However, bio-based production faces a challenge given the economies of scale in petro chemistry. Genomics applications have a great potential for bio-based sustainable economy and development.

Renewable aromatic material lignin, a complex organic polymer deposited in the cell walls of many plants, making them rigid and woody, which creates the challenges of availability in the biosphere exceeding 300 billion tons and increasing annually by around 20 billion tons (Smolarski, 2012). Other major challenge is the bio economy based on genetic engineering for strain improvement and higher biomass and green product yields, and the need to gain market and regulatory acceptance of such organisms (Sayre et al., 2013).

Sustainable bio economy is more than environmental sustainability, rural and urban industrial ecosystems, sustainable city development, wealth creation, green production, etc. It is a concept related to green rural and urban revitalization and regeneration, smart city, green re-industrialization, bio production, biodiesel refineries, creation of value chains.

SEARCHING FOR A SOLUTION

To tackle the increasing climate change and socio-economic problems in urban settlements is required to incorporate sustainable urban spatial planning in sustainable city development under the term sustainable city development. Among the functions of sustainable city development is planning and projecting the population distribution and land regulation under sustainable development criteria in cities and metropolitan areas (SEDATU, 2015).

The effectiveness of the sustainable urban planning and development is linked to the size and share of the informal city, development regulations and availability of resources. Sustainable city planning and development legislation and open dialogue with stakeholder groups are important factors for the urban development of green urban areas, sustainable city development and urban containment and densification. Sustainable urban planning is a governance tool to achieve sustainable city development. Urban planning is a core component of an urban governance model to promote democracy, inclusion, participation, transparency and accountability aimed to ensure sustainable city development.

Urban planning needs an integrated approach to cover urban policy, governance and sustainable city development. A holistic approach to sustainable city development integrates sustainability through urban planning. Sustainable strategic city development planning focuses in developing a common vision, mission, goals, strategies and policies, as the result of stakeholder´s risk sharing. The urban development planner must sustain his behavior in ethical grounds and moral philosophy framed by a neo modernist sustainable city development and urban planning ethos.

Bio economy is part of the solution to face the challenges of a sustainable city development if it can be economically growing, environmentally sustainable and socially inclusive and equality oriented. The benefits coming out from bio economy are notwithstanding because there are several defining challenges that are proving difficult to overcome (Cheali et al., 2015). Bio economy refers to the activities spread across diverse economic sectors relating to the invention, creation, development, production, distribution, consumption and use of biological products and processes.

The bio economy is a discipline that considers new green technologies that are designed and developed as biotechnological processes and evaluated their sustainability and performance in response to social challenges such as green food safety, smart cities, waste disposal, green urban areas, sustainable energy, health, digital and networking security, etc. Bio economy research and innovation is conducted in specific programs related to food security, urban green innovation areas, sustainable agriculture and forestry, marine, maritime and inland water.

Bio economy strategies in sustainable city development are formulated and implemented to overcome the challenge of sustainability on many levels. Bio economy policy ensures biomass sustainability by collecting and using it within the boundaries of natural resources that the planet can provide bio economy policy frameworks and environmental ecosystems can tackle some of the toughest of the grand challenges such as climate change, food security energy security and resource depletion. The diversity of economic sectors includes agriculture, fishery, forestry, food, waste management, health, trade, etc.

The bio economy concept emerged and has grown from a biotechnology-centric vision to be applied in economic activities giving place to a more complex bio economy policy framework and sustainable city development and environmental ecosystems. Despite the challenges, the biotechnology production offers several advantages over the chemical production. The chemical industry poses challenges largely invisible in manufactured goods despite the essential role it plays. The traditional chemical industry facing the challenge of struggling with a poor public image (Moreau, 2005) is being replaced with new bio-based materials in a range of different bio-based chemicals already close to the market.

The first-generation of biofuels and bioenergy had lower added value and job creation potential. Incineration with energy capture can be used to tackle the waste plastics problem (Yamashita & Matsumoto, 2014). A new agenda centered on strategic innovation and research aims to have added value increasing the wealth creation and employment and high economic growth by meeting the grand challenges posed by sustainable rural and urban regeneration, energy security, climate change, textile, food processing, chemicals industries, health industries, etc.

The ongoing research and innovation in bio economy production is rising in some industrial sectors such as the chemical, textile and other industries and their contributions to sustainable city development. However, there is an urgent need of market research, studies, analysis and development. Biggest changes are taking place in bio economy markets for the last decade.

ORGANIZATION OF THE BOOK

The content of this book aims to reflect and analyses the innovations currently taking place in the changing bio economics policy framework and environment and the attraction of investments to propose and design solutions and its implications in sustainable city development.

The book is divided in three sections. Section 1 is dedicated to the chapters dealing with the conceptual and theoretical bio economy framework of reference. Section 2 aims to concentrate more on issues related to investment in different industry sectors. Section 3 presents some specific cases on sustainable city development.

SECTION 1: CONCEPTUAL AND THEORETHICAL BIO ECONOMY FRAMEWORK OF REFERENCE

Chapter 1 – Bio-Economy: Visions, Strategies, and Policies

Bio economy is transforming the world`s economy from conventional to sustainable mode. This chapter analyzes the development of sustainable bio economy through the implementation of the adequate visions, key strategies and policies in some specific cases.

Chapter 2 – Strategies of Sustainable Bio Economy in the Industry 4.0 Framework for Inclusive and Social Prosperity

The chapter makes a descriptive analysis to the unpredictable transformations in knowledge related areas in what has been called the fourth industrial revolution, basically a knowledge revolution that have affected many areas of knowledge, among them the bio economy. The chapter offers some bio economic strategies to further the Industry 4.0 relations.

Chapter 3 – Urban Farming in Sustainable City Development

This chapter deals with urban Farming by integrating urban space and buildings in green spatial systems for the preservation of the environmental balance of urban spaces and connections concerning the coexistence of architecture and greenery in the city.

Chapter 4 – Parametric Evaluation of Beam Deflection on Piezoelectric Material Using Implicit and Explicit Method Simulations: A Study in Energy Engineering

The chapter deals with an energy harnessing method from rainwater using the piezoelectric materials with the purpose to visualize an interconnection in parametric analysis of piezoelectric effect-based energy harvester using two different commercially available piezoelectric materials

SECTION 2: ISSUES RELATED TO INVESTMENT IN DIFFERENT INDUSTRY SECTORS

Chapter 5 – Entrepreneurial and Institutional Analysis of Biodiesel Companies in Mexico

This chapter makes an institutional analysis of entrepreneurial activities in biodiesel companies and their important contribution to the bio economy energy sector in Mexico and its relationship with quality and maturity of joint relationships. The analysis concludes that the institutions in Mexico have achieved their maturity and incentive to increase the number of producers and distributors of biodiesel thus taking advantage of the growing market.

SECTION 3: SPECIFIC CASES ON SUSTAINABLE CITY DEVELOPMENT

Chapter 6 – Ecological Centre of Warsaw as a Development Path

The purpose of this chapter is to present a new development path towards greening the city center of Warsaw by converting disperse green areas into a continuous network of green areas in the context of the increasing density of building development on both sides of the Vistula River. This future structure of green areas is based on the elements of bio economy.

Chapter 7 – Green Spaces of the Metropolitan Area of Guadalajara

This chapter analyses the green areas and public parks of the Metropolitan Area of Guadalajara as a system that shares common problems with other Latin-American cities, such as the deficit of urban green spaces, insecurity, unemployment, social exclusion and stress, etc.

Chapter 8 – Application of UAV Technology to Planning Study on Chinese Villages in Guanzhong

The chapter describes the multiple problems that the Guanzhong region in China faces in its rural planning research to implement the Chinese state strategy of rural revitalization. Despite this situation, this rural area offers better work environment conditions to implement UAV technology

Chapter 9 – Developing a Sustainable Eco-City in Pre-Olympic Tokyo: Potential of New Methods and Their Limits in an Urban Era

This chapter deals with the emergence of sustainable urban development in Tokyo in light of the upcoming 2020 Olympics by exploring some practices such as rooftop- and wall greening or river projects in the inner city.

Chapter 10 – Forest-River-Ocean Nexus-Based Education for Community Development: Aiming at Resilient Sustainable Society

This chapter elucidates bio economic research aimed to developing a more resource-efficient and sustainable society that uses renewable biological resources to produce food, materials, and energy. This paper examines the effectiveness of forest-river-ocean nexus-based education for community development (FRONE) in encouraging the sustainable use of biological resources.

CONCLUSION

This book has a positive impact and makes significant academic and scientific contributions to the subject matter of bio economy solutions and investment to sustainable city development by such a diverse group of scholars from around the world. These contributions further advance and deepen the theoretical and conceptual frameworks of the elements and attributes to bio economy as a new scientific approach to sustain the practice of sustainable city development projects, which will be crucial in future. As a concept, the relevance of bio economy encompasses economic growth and efficiency driven by the sustainable development of renewable biological and natural resources and biotechnologies to produce sustainable products and services, creation of value, employment and generation of increasing income.

Sustainable city development is under pressure for more efficient economy, more inclusive societies, more sustainable environment based on more rational use of natural resources to meet competing and complementary objectives in energy, health, food, water and other important urban sectors interrelated with sustainable urban green areas, housing, building, public spaces, etc. of urban environments. The interactions and interrelationships between these sectors have become increasingly affected by the investments in bio economy.

This book explores the bio economy and its interlinked relationships with investments and sustainable city development elements as three relevant interacting variables that are essential requirements aimed to influence affecting and impacting the execution of urban projects for developing more livable sustainable sites. The bio economy based-strategies, policies and tools described and explained in each chapter, either at the theoretical, conceptual or methodological level of analysis or in the specific cases, demonstrate the intention and commitment the implications in real life. Besides the direct implications the chapters of this book has in sustainable city development, it is wise to recall other implications to the large range of related policy formulation and implementation in other areas such as urban building and housing, environmental and ecological development, urban innovation, business and industry operations, trade, urban farming, urban green areas, taxes, waste materials, etc.

At least but not the last, all the authors involved in writing the chapters are academic experts in their fields who are elaborating their proposals for sustainable city development based on investments and bio economy. The implementation of their proposals are dependent on sustainability of the feedstock and the processes of the bio economic products and services to turn around the negative trends in global warming, pollution, and in general the unsustainable city development implemented until now.

Let us analyze, follow and implement the advices of the authors of the chapters of this book to work for a healthier and ecological urban environment, more economic efficiency and more equal and inclusive city, because the mistakes of the past are not to be repeated in the future. Let us design a better place where to live with human dignity.

José G. Vargas-Hernández
University of Guadalajara, Mexico

Justyna Zdunek-Wielgołaska
Warsaw University of Technology, Poland

REFERENCES

Cheali, P. (2015). Upgrading of lignocellulosic biorefinery to value added chemicals: Sustainability and economics of bioethanol-derivatives. In Biomass & Bioenergy (Vol. 75, pp. 282–300). Amsterdam: Elsevier.

Moreau, N. J. (2005). Public images of chemistry. *Chemistry International*, *27*(4).

OECD. (2018). *Meeting Policy Challenges for a Sustainable Bioeconomy*. Paris: OECD Publishing. doi:10.1787/9789264292345-

Sayre, R. (2013). Initial risk assessment of genetically modified (GM) algae for commodity-scale cultivation. Algal Research, 2(1), 66-77.

SEDATU. (2015). Available: http://www.sedatu.gob.mx/sraweb/

Smolarski, N. (2012). *High-value opportunities for lignin: Unlocking its potential.* Frost & Sullivan. Retrieved from www.greenmaterials.fr/wp-content/uploads/2013/01/High-value- Opportunities-for-Lignin-Unlocking-its-Potential-Market-Insights.pdf

UN. (2009). *Human Settlements Program. In Planning Sustainable Cities: Global Report on Human Settlements 2009*. London: Earthscan.

Yamashita & Matsumoto. (2014). Status of recycling plastic bottles in Japan and a comparison of the energy costs of different recycling methods. *International Journal of Environmental Protection and Policy*, *2*(4), 132-137.

Section 1

Conceptual and Theoretical Bio-Economy Framework or Reference

This section presents four chapters addressing some relevant issues focusing on concepts and theories of the bio-economy paradigm. The four chapters conceptually explore some different trajectories and contestations of this paradigm, arguing more broadly that this is becoming a significant driver for re-structuring corporate and competitive strategies and policies between firms, markets, and urban and rural spatial development between and within places.

Chapter 1
Bio-Economy:
Visions, Strategies, and Policies

Baseem Khan
https://orcid.org/0000-0002-0562-0933
Hawassa University, Ethiopia

ABSTRACT

As the world's economy is transforming from conventional to sustainable mode, bio-economy will play an important role in this transition. The development of sustainable bio-economy is the great challenge in different parts of the world. For this purpose, adequate visions, strategies, and policies are required. Therefore, this chapter deals with the adequate visions, key strategies, and important policies for the development of sustainable bio-economy. Some case studies also presented where these visions, policies, and strategies are already implemented.

DOI: 10.4018/978-1-5225-7958-8.ch001

INTRODUCTION

In the different parts of the world, for sustainable and smart development (Fanuel et al. 2018), conventional fuel based economy is transforming in to bio based economy model, with novel innovations in life science and bio chemistry. These innovations are prompting the development of new visions, policies and strategies in bio based economy model. The concept of bio-economy is evolved from the ever increasing research and innovations in the field of bio-based energy sources, bio-materials and chemicals. In several countries, strategies and policies, related to bio technology and bio-based industries and products are already developed. Further, in many other developing nations, above mentioned topics are discussed under the category of bio-economy. Various aspects of economy, environment and society will be affected by the advanced and larger bio-economy. A country can be developed strategies for bio economy, to present its interest towards conservation of bio diversity, protection of rare biotopes, preserve amount and quality of food and climate change. Strategies, policies and visions for bio-economy are developed and adopted by the various countries around the world. It is highlighted the fact that these are very important for sustainable development (OECD, 2009; EC, 2012). Publication of "the organization for economic cooperation and development (OECD)", entitled "The Bio-economy to 2030: Designing a Policy Agenda (OECD, 2009)" stated that many resources and health related issues, which are facing by the world can be solved by using the solutions, provided by the biological sciences. Further, biomass based energy production scheme can also be incorporated in recent micro-grid technologies (Khan and Singh 2017; Singh and Khan, 2017).

These solutions can be provided a way to enhance sustainability in economies, although, research findings and innovations are the basis of defining policies that are related to the bio-economy. According to OECD, both private and public sectors must be taken into account to design any policy matter, for maximizing the impact and potential of bio-economy.

This chapter is organized in the following manner. Section two discussed the background of the bio-economy; third section discussed three visions associated with bio-economy. Various bio-economy strategies are presented in section four. Section five discussed the bio-economy policies and policy mix for sustainable bio-economy development, which is followed by the conclusion.

BACKGROUND

The main aim of this chapter is to address various visions, strategies and polices, which are utilized by various countries for the development of sustainable bio-economy.

Though, to develop a strategic framework, the OECD agenda on bio-economy policy and strategy development is required to discuss first. The criteria for the selection of countries and regions to develop sustainable bio-economy is that these regions should rich in bio based resources and they should have research experience in the area of bio-economy development. Regions should have large forest areas that will be main aspects for the development of bio-based economy. Along with that regions should have innovations and research in the field of bio-products based industries and bio-refineries.

For example Germany, which is an important nation in European Union, announced the closure of all nuclear power plants. This decision will enhance the renewable energy generation and bio-economy development. Further, Australia is also facing problems of water scarcity and bad impacts of climate change. Therefore, Australia is also increasing its renewable energy production and bio-economy development. Significant bio-economy potential is available in the above discussed both the countries. European Union as key players in the field of bio-economy pays more attention on knowledge of bio-economy, which includes research, development and presentation of various projects in the fields of bio-energy, agriculture and bio-refineries. Large number of scientific articles has been published in the field of bio-economy. Bio-economy concepts present a way by which bio-resources can be utilized in the economy. OECD defined the concept of bio-economy as "transforming life science knowledge into new, sustainable, eco-efficient and competitive products (OECD, 2009). According to OECD (OECD, 2006), innovations have great potential to utilize bio-based products in more efficient way. Leading nations such USA also defined the notion of bio-economy as follows: "A bio economy is based on the innovation and research in the field of bio-science to develop economic activity and public benefit (WH, 2012).

Basically, bio-economy incorporates the methods of conversion, which converts raw materials into useful products.

BIO-ECONOMY VISIONS

This section presents the various concepts of bio-economy around the world by considering different origins and dispersion of bio-economy ideas across numerous scientific fields. There are three ideal types of visions, which can be incorporated by the concept of bio-economy (Levidow et al., 2013; Staffas et al., 2013). In the field of natural and engineering sciences, the importance of research in the field of bio-economy can be reflected by using the first two visions, which are considerably affected by a technical perspective. These visions are as follows:

1. **Bio-Technology Vision:** It emphasis on the importance of the research, application and commercialization in the field of bio-technology for various sectors.
2. **Bio-Resource Vision:** It focuses on the research, development and presentation activities in the field of biological raw products, for various sectors such as marine, agriculture, bio-energy and forestry. Furthermore, it is associated with the establishment of new value chains.
3. **Bio-Ecology Vision:** It highlights the significance of various bio-ecological processes, which optimize the utilization of nutrients and energy, support bio-diversity and avoid degradation of soil.

Bio-technology vision mainly incorporates applicability of science, while bio-source vision emphasis on the conversion and up-gradation of biological raw materials. These two visions (Bio-technology and bio-source) are mainly focused on technology and provided the main role to research, development and presentation. On the other hand, bio-ecology vision emphasis on integrated and circular processes and systems.

These visions are not completely separated to each other, but these all are ideal visions for bio-economy. Therefore, when the major organizations are associated with different visions, presented in table 1, than these visions are interrelated to each other (Levidow et al., 2013; Staffas et al., 2013).

In European commission, the initial policy making is greatly affected by the research work on the bio-technology vision (Richardson, 2012). Features of three bio-economy visions, which focusing on various implications such as overall aims, objectives, drivers, value creation, mediators of innovations and spatial focuses are discussed in the following sections.

Bio-Technology Vision

The main aims and objectives of the bio-technology vision are related to the job creation and economic development (Staffas et al., 2013; Pollack, 2012). Therefore,

Table 1. Different organizations and visions

Organizations	Vision
OECD	The bio-technology vision
The European Commission	The bio-resource vision
The European Technology Platform (TP) Organics	The bio-ecology vision

economic development is prioritized over sustainability, if positive impacts of environmental aspects and climate change are assumed. Thus, the feedback effects are normally ignored in the utilization of bio-technology vision (Richardson, 2012). Other priorities such as ethical concerns and risks are secondary for economic growth (Hilgartner, 2007). Applications of bio-technologies are linked to the job value creation in various sectors. Further, research and developments' commercialization are also related to the job value creation. Capitalization of bio-technologies will follow the economic development. Further, for simulating economic development around bio-economy, intermediaries between investors and research firm played a vital role (Morrison and Cornips, 2012). Therefore, the main aspect in bio-economy is the investment in the research and innovation that will result in the generation of scientific knowledge. Further, research is initiated from processes those are operated at the level of molecule. Consequently, products and production processes are constructed. The above discussed procedure transforms bio-mass into useful market products (Hansen et al., 2014). In bio-technology vision, the inherent understanding of innovation processes, which are related to mediators and drivers, is similar to linear model of innovation. In that model, innovation is assumed to be initiated with scientific research. Further, this innovation is tracked by development, fabrication and marketing (Hansen and Winther, 2011). Among industries and universities, a close interaction is required in the innovation process for ensuring commercialization of relevant research (Zilberman et al., 2014). Resources scarcity can be removed with the help of technological process in the bio-technology vision.

Therefore, shortage of resources is not the main parameter of analysis (McCormick and Kautto, 2013; Staffas et al., 2013). As bio-technology processes result in very less or no amount of waste, management of waste is not a key problem in the bio-technology vision. These processes are designed at the molecular level to produce very less amount of waste. The new end products can also be obtained from the organic wastes by bio-technologies (Richardson, 2012). Once the bio-technologies reached to commercial level, these technologies have wide range of applications that can reduce the boundaries between traditional production approaches (Wield et al., 2013; Boehlje and Boehlje, 2011). From the above discussion, it is clear that research is the most important part in this vision; therefore, research funding organizations and research councils are the main agents to translate bio-technology vision into the real development in the area (Kearnes, 2013). Other than research, issues of governance such as policies for bio-economy also has prominent role in this vision (Aguilar et al., 2013).

As a spatial point of view, this vision of bio-economy concentrated the growth in a limited number of regions, which host the huge pharmaceutical industries with small bio-technology organization and venture capital (Cooke, 2007; Cooke, 2009). Moreover, regions with specialized public research in bio-technology may also get advantages in terms of development (Birch, 2009). These global bio-economy centres should be connected to each other for bio-economy's innovation. This will help emerging regions to develop their economies by taking advantage of innovations in bio-economies. From the global competition point of view, innovation in governance is a key idea in some of the research planning in bio-economy vision (Rosemann, 2014). Value creation associated with the geographies in the bio-economy consists of different material component of bio-resources. Immaterial components such as knowledge and capacity, to develop new knowledge, are also associated with the value creation in bio-economies. The other major issues include strategies and conditions, applied to build a bio-economy in emerging regions (Birch, 2012; Chen and Gottweis, 2013).

Bio-Resource Vision

In this vision, the objectives and aims are related to both sustainability and economic growth. The main expectation from the bio-innovation in this vision is to deliver both environmental sustainability and economic development (Levidow et al., 2013). In bio-technology vision, economic growth would be followed by capitalizing on bio-technologies, whereas in bio-source vision, economic growth would be followed by capitalizing on bio-sources. As it is believed that the environmental sustainability impacts are positive, instead of giving attention on environmental protection, the major focus is provided on development of technology for new bio-dependent products. Therefore, the effect of climate change on the transition of bio-economy isn't frequently assessed. Due to that, policy makers provide limited attention on the sustainability aspects (Ollikainen, 2014). Despite questioning the positive sustainability impacts of bio-economy by various academicians, policy makers weakly integrate the policies of bio-economies. Ponte (2009) argued that instead of focusing on the outcomes in the form of sustainability, procedures and processes related to standard tuning in the bio-economy are more important. Less emphasis is provided on the concerns such as deforestation and depilation of bio-diversity, in the discussion of bio economy (Pülzl and Kleinschmit, 2014). This vision highlights the transformation and processing of bio-resources into new value added products, in terms of value creation. Based on the utilization and accessibility of bio resources, waste management also requires more attention in bio-resource vision. The main concern in the value chain is the minimization of the production of organic waste. Further, this waste production is the vital input for the generation of renewable energy (EC, 2012). Cascaded utilization of biomass is the central idea in this regard to highlight the efforts of maximizing the efficacy of biomass utilization (Keegan et al., 2013). For large scale bio-fuel production, those processes are very important, which permit recycling of the waste material by converting it into fertilizers. In term of agents and drivers of innovations along with the natural outcomes of the main idea of bio-resources, the matter of land utilization consists of clear cut component as comparison of bio-technology vision. Therefore, the important factors in bio-resource vision are the enhancement of land productivity and incorporation of degraded land for the generation of bio-fuel (Mathews, 2009). Though, very less discussion is performed on the implications of other aspects such as climate change, on exchanges between various types of land utilization (e.g. agriculture and forestry). Moreover, the case, when concerns related to the availability and utilization of the bio-resources are important, and then the connection between usage of other resources (i.e. fertilizers, water, etc.) and bio-resources are rarely analyzed (Staffas et al., 2013). Bio-resource vision also pointed out the importance of innovation and research for value production in a similar way to bio-technology vision. Basically, bio-resource vision emphasized on the significance of innovation and research in several fields those are connected to biological materials in different ways. On the other hand, bio-technology vision has a very narrow point of variation in the research of bio-technology. Accordingly, the collaboration between various agents with different capabilities is often related to the practices in research and innovation. The significance of research in the areas for example consumer's preferences is also focused (Levidow et al., 2013). For developing collaboration between different sectors, innovation is very important. For example, organisations from the forestry sectors are closely associated with downstream agents. The significance of inter-sectoral associations is also regularly highlighted in the policies for bio-economy innovations (McCormick, 2013). Therefore, in the bio-source vision, innovation factors like relations with consumers and cross-sectional collaborations, which highlight value creation, are less linear as compare to bio-technology vision, are focused. In the context of spatial focus, this vision focuses the considerable potential to fuel growth in rural areas. It is discussed that trees and plants are producing new bio-products, which will optimistically affects the job creation in rural areas. Further, this will be less mobile in nature as compare to other economic activities because of the significance of natural resources at important locations (Low and Isserman, 2009). Therefore, bio-economy, based on the bio-resource vision is available for the revitalized growth of rural areas through variation into value added products. As the localized capabilities of farming and processing bio-mass are main idea of growth, it will be supplemented with huge localized information (Albert, 2007).

Bio-Ecology Vision

Sustainability is the main concerned in the aims and objectives of the bio-ecology vision. As employment generation and economic development are the main issues in the bio-resource and bio-technology vision, respectively, these features are less important than the sustainability aspects in the bio-ecology vision (Levidow et al., 2013). The available research on bio-economy incorporated the issues and critical points, which are focusing on economic development and commercialization in the bio-technology and bio-resource vision. It also reflects the attention and issues related to the sustainability. Several literatures in the area of health science disapproved the commercialization of bio-resources in such areas, where human tissues like foetal tissue, stem cells and blood are considered (Mumtaz et al. 2012). These are the areas, which require ethics in the commercialization of bio-resources (Bahadur and Morrison, 2010). Bio-ecology vision focused on the following issues for value creation (Levidow et al. 2013; McCormick, 2013):

1. Encouragement of bio-diversity
2. Preservation of Ecosystems
3. Capabilities to supply ecosystem services
4. Conservation of soil

Further, it is focused that bio-waste produced energy only after the end of supply chain following reuse and recycling. For bio product generation industries, the use of own or urban waste is very important to reduce or eliminate the requirement of input from external resources. Therefore, this vision mainly focuses on self sustained creation mode. The bio-ecology vision points out the recognition of suitable bio-ecological exercises, in the context of highlighted factors and agents of innovations (Marsden, 2012; Siegmeier and Möller, 2013). Further, it also focused the ecological interactions associated with the recycling and re-utilization of waste bio-mass and its efficiency in land utilization. Another important area is bio-ecological engineering techniques, which focus on the designing of agricultural schemes that need very less agrochemicals and low energy input instead of depending on various ecological interfaces among different biological elements for enabling agricultural schemes to enhance their crop production and security and soil fecundity. As seen that previous two visions are mainly concerned with application of technically supported innovation and research actions, but this approach is not true in the case of bio-ecology vision. Indeed some hybrid techniques like genetically modified corps is not considered in this vision. It doesn't show that innovation and research technologies are not important for this vision but rather theses techniques have other focus (Albrecht et al. 2012). For example, some researchers focused on trans-disciplinary sustainability issues, which are related to the ethical global trade, sustainable biomass cultivation potentials, dialogues and resolutions on different transition procedures. At last, research considered the negative impacts of competing different bio-economy visions. In terms of geographical manner, this vision focused on the different prospects for peripheral and rural areas, in the same manner to the pervious bio-resource vision. High value products with territorial identity may results in rural development opportunities. Though, the significance of peripheral relations focused in bio-resource vision, bio-ecology vision described for the growth of locally surrounded economies as the main element of the attempts for ensuring a sustainable bio economy (Marsden, 2012).

BIO-ECONOMY STRATEGIES

There are several steps, which are incorporated into the research of bio-economy strategies. In the first step, sampling and selection of the strategies are performed. This section deals with various existing fully integrated bio-economy strategies around the world. Germany, EU and other international bio-economy strategies are discussed in this section. In the next step, re-construction of the different processes is performed, which guide towards the relevant strategy. For that purpose three characteristics are discussed. These are as follows:

1. Variations in the focus and understanding of the bio-economy
2. Linkages to different governmental policies and strategies
3. Members associated in the process of drafting

Therefore, different agents and perspectives are analyzed. After that the assessment and classification of various bio-economy definitions used in different strategies are conducted. System boundaries are affected by the different understanding of bio-economies, for example the types of economic activities considered as the component of bio-economy, which affects the economic assessment (Efken et al., 2016). Basically, bio-economy definition has different implications based on the activity of areas. For innovation and transformation procedures and their control, socio-technical visions have turn into an ever increasing component. The above discussion is true for complicated and comprehensively new and promising techniques such as nano-technology. Further, it is also true for the transformation of already existing huge socio-technical systems like energy transition system (Lösch and Schneider, 2016). For addressing, communication and coordination between different agents in the innovation and research, visions are performing as instruments. These visions also affect the social spheres and groups by endorsing novel objectives and elucidations for different social problems. Visions also provide target innovations as fundamental innovations (Brown et al. 2000). Further, visions also participate as the vital element for developing bio-economy strategies. The assessment of visions is initiated from the definite group of promises and prospects in every strategy, by which different vision are developed and classified.

For the implementation of strategies and the road of development for bio-economy, the following step is analyzed the guidelines and principles. Prerequisites and requirements are formulated by this step. For instance the principle of sustainability has developed in most of the policy areas. Guidelines and principles play an important role for the biomass utilization in the context of bio-economies strategies. The evaluation of the association among different scientific and social issues and strategies around bio-economy is based on the various published literature.

Analysis of Bio-Economy Strategies

Bio-Economy Strategies: Back Ground and Formation

The notion of bio-economy is comes from the spheres of bio-technology and life sciences (McCormick, 2013). Different policy making approaches (i.e. national and international) are less or more complicated and leading to bio-economy strategies, with varying perspectives. This chapter provides a comprehensive impression of the different processes and their relationship with governmental policies for developing bio economy strategies.

Bio-Economy Strategies: International and European

OECD report "the Bio-economy to 2030: designing a policy agenda" (OECD, 2009) is an insight project that investigated the future of bio-economies in long-term. This report is supervised by different representatives from various organisations such as governments, international organizations and industries and is supported by various experts from industry and academics. The secretariat of OECD drafted this report. In 1982, a conman research policy for funding is started in EU, which is provided for the research program on bio-technology (Bio-molecular engineering program).

In the next three decades, several bio-technology research programs are started (Aguilar et al. 2013). A knowledge based bio-economy program is started in 2005 by the European commission. The source of word "knowledge based" is from Lisbon strategy that stated that until 2010, EU will be the most vibrant and competitive knowledge based economy in the world (EC 2000). In 2007, the famous Cologne paper is presented. Experts from research and industries drafted this paper, which focused on the bio-technology outlooks until the year 2030 (GP 2007). In 2010, under the Belgian Presidency, a report is presented in the conference entitled "the knowledge based bio-economy towards 2020" (EC, 2012). This report presented the achieved goals and exceptional challenges (BP 2010). This report abandoned the focus on bio-technology and addressed the new areas of bio-economy such as common agriculture policy, and sustainability criteria. In 2011, the transition from bio-technology to social issues is presented in the white paper entitled "the European bio economy in 2030" (BECOTEPS, 2011). This paper presented the vision for sustainable, inclusive and smart bio-economy for EU. This paper is the outcome of discussion for various experts from nine different technology platforms; which covers the wide range of features of bio-economy. EU bio-economy strategy is approved at the end of this process, which is associated with the commission staff working documents (EC, 2012) with detailed background and action plan. This bio-economy strategy is incorporated in the complete strategy i.e. "Europe 2020" (EC 2010). The two main initiatives "A resources efficient Europe" and "Innovation Union" should be contributed by bio-economies of "the Europe 2020 bio-economy strategy" (Scarlat, 2015). A public consultation meeting is organized by the European Union for the preparation of the bio-economy strategy. Over 200 responses are received by EU for bio-economy strategy development. The vital elements for the implementation of bio-economy strategy are innovation and research initiatives i.e. "Horizon 2020", which are available for duration of 2014 to 2020. In this, bio-technology, food security, sustainable agriculture marine, water research and bio-economy provided the specific budget of 4.1 billion Euros for the period of 2014-2020 (EC, 2012).

Bio-Economy Strategies: German Perspective

In 2009, with the appointment of the bio-economy council for the duration of three years, by federal ministries, the policy development on bio-economy is started in Germany. For the incorporation of broader areas, a new council is appointed in 2012 that represented the areas of science, economy and society. But lately, only social scientist covered the council. In 2009 (Bioökonomierat, 2009), first recommendations by the bio-economy council is published with spotlight on research organization.

A detailed report is published after one year in the form of publication of short strategic paper (Bioökonomierat, 2010). External experts provided the support to the council for drafting this report. Three broad areas are covered by this detailed report along with programming recommendations. The "National research strategy Bio-economy 2030" is released in the same year 2010 (BMBF, 2010). The federal ministry for education and research in support of six other ministries drafted this report. There are various similarities in the recommendations of this report to the recommendation of bio-economy council. An overall funding volume of 2.4 billion Euros is assigned till 2016 for the development of strategy framework. The high-tech strategy of federal government incorporated this bio-economy strategy for initiating inter-ministry research initiatives and is organized on highest priority to social issues. In 2013, the "national policy strategy bio-economy" is approved by federal government. It is developed by the federal ministry of food and agriculture with the support of bio-economy council and other four ministries. By discussing different areas, its structure is different from the research strategy. This policy strategy is incorporated at various strategic levels in federal government (BMEL, 2014).

In addition to that, individual bio-economy strategies are developed by the two federal states of Germany. The ministry of innovation, science and research of federal state of North Rhine-Westphalia developed its own bio-economy strategy (Capgemini Consulting, 2010). Similarly, the working group of scientists developed a bio-economy strategy for the ministry of science, research and art of federal state Baden-Württemberg. This strategy has a broad range of disciplines related to bio-economy. It includes wide range of current research landscape of this federal state. It also incorporated the SWOT analysis with reasoning and description of topics to be funded (MWK BW, 2013).

Bio Economic Strategies: Other Countries Perspective

In 2010, research and innovation strategy for bio economy is launched in Sweden (FORMAS, 2012). The strategy is developed by the Swedish Council for Environment, Agricultural Sciences and Spatial Planning in cooperation with Swedish Governmental Agency for Innovation Systems and the Swedish energy agency on the request of Swedish government. A detailed discussion between the members of research groups, industries and companies are performed at the time of drafting phase. In Sweden, bio-economy activities distributed between different organizations and institutions (Winther, 2016).

The 2012 national bio-economy strategy of USA focused on administrative steps to harness research innovations in bio-economy. The office of science and technology policy and executive office of the president with the cooperation of various federal departments drafted this strategy. A public consultation is also started before half a year. More than 135 responses are received from institutions and individuals mainly from the field of science and technology (OPST, 2011).

POLICIES FOR SUSTAINABLE BIO-ECONOMY DEVELOPMENT

Basis for Bio-Economy Policies

Bio-economy polices are not restricted to the area of forestry, agriculture, aquaculture and fishery. These are not sector specific policies. An essential element for the sustainability of entire society is the transformation of economy towards sustainable bio-economy. It requires transition procedures across the whole economy. For instance on production side, by investment in research infrastructure and manifestation activities, whereas on consumption side, by changing policies that relates to investment and consumption patterns and through public procurement measures. Particularly, for guiding the growth of bio-dependent end products, sustainable public procurement models can be the main force as it considers major part of GDP. Thus, policy is the main force, which decides the way of transformation process, going ahead as compare to conventional market failure decision policy and various specialized strategies. Conventional market place failure strategies are differentiated by Weber and Rohracher (2012) as information asymmetries, huge exploitation of commons, and externalization of cost and information spill over. Klein et al. (2005) classified system failure into two ways. First, structural system failure are identified as failure of institution, network failure's interaction, failure of capability and infrastructure, whereas transformational system failure are the failure of demand articulation, failure of directionality, failure of policy coordination and failure of reflexivity. These different rationales of current policies must be incorporated in to the policy for bio-economy. A clear differentiation must be existed between different countries on the understanding of bio-economy.

These bio-economy policies must envision commercialization and application of bio-technology vision in different industrial areas. Their aim must be to process bio-based resources as main driving factor and aims of innovation and economic development. Further, these policies must highlight the significance of ecological process, which optimize the utilization of nutrients and energy, support bio-diversity and prevent soil degradation. These policies should provide more attention to social and environmental aspects in the utilization, by providing support to private consumers for making more sustainable options. The policy should incorporate the different bodies like civil society, industries, businesses, state and municipalities. Further, by considering digitalization, a digital strategy can also be developed, which follows the sustainable digital transformation vision. The general objective of bio-economy policies is to harness various prospects of bio-economy visions.

Transition to Sustainable Bio-Economy by Policy Mix

In many experimental studies, technological innovation system technique is demonstrated. Presently main focus will be provided on dynamics within an emerging technological innovation system. The concept of "functions of innovation system" is developed to address the dynamics, by selecting key activities. The most vital processes are focused, which require performing in innovation system that leads to technology diffusion and growth (Hekkert et al. 2007). Functions like knowledge development, guidance of search, formation of market and resource mobilization are important features in the policy mix. For system change, policy mix approaches are analyzed for combining framework of technological innovation and regime destabilization.

CONCLUSION

For sustainable city development, bio-economy is the main element. With the help of bio-economy, a sustainable development infrastructure can be developed. Therefore, critical issues related to the bio-economy development such as visions, strategies and polices are discussed in this chapter. Three basic bio-economy visions such as bio-technology vision, bio-resource vision and bio-ecology vision are discussed in detail. Then, various bio-economy strategies, which are already implemented in different regions such as European Union, Germany, USA and Sweden, are discussed. At last bio-economy policies with policy mix are presented with policy for destabilization. This chapter will be helpful for developing regions for the development of bio-economy in their regions. This literature will take as a reference for sustainable bio-economy development as it consists of visions, strategies and policies.

ACKNOWLEDGMENT

The authors would like to thank all of colleagues for their support and cooperation. Authors would also like to thank Department of Electrical and Computer Engineering, Hawassa University for providing the environment and support to carry out this work.

REFERENCES

Aguilar, A., Magnien, E., & Thomas, D. (2013). Thirty years of European biotechnology programmes: From biomolecular engineering to the bioeconomy. *New Biotechnology*, *30*(5), 410–425. doi:10.1016/j.nbt.2012.11.014 PMID:23195849

Albert, S. (2007). Transition to a bio-economy: A community development strategy discussion. *Journal of Rural and Community Development*, *2*, 64–83.

Albrecht, S., Gottschick, M., Schorling, M., & Stint, S. (2012). Bio-economy at a crossroads. Way forward to sustainable production and consumption or industrialization of biomass? *GAIA Ecol. Perspect. Sci. Soc.*, *21*, 33–37.

Bahadur, G., & Morrison, M. (2010). Patenting human pluripotent cells: Balancing commercial, academic and ethical interests. *Human Reproduction (Oxford, England)*, *25*(1), 14–21. doi:10.1093/humrep/dep369 PMID:19897856

BECOTEPS (Bio-Economy Technology Platforms). The European Bioeconomy in 2030. Delivering Sustainable Growth by Addressing the Grand Societal Challenges. (2011). BECOTEPS. Available online: http://www.epsoweb.org/file/560

Belgian Presidency. (2010). *The Knowledge-Based Bio-Economy (KBBE) in Europe: Achievements and Challenges*. Belgian Presidency (BP). Available online: http://www.mercadosbiotecnologicos.com/ documents/the_knowledge_based_bioeconomy_kbbe_in_europe.pdf

Bioökonomierat (German Bio-economy Research and Technology Council). (2009). *Combine Disciplines, Improve Parameters, Seek out International Partnerships. First Recommendations for Research into the Bio-Economy in Germany*. Forschungs- und Technologierat Bioökonomie. Available online: http://biooekonomierat.de/fileadmin/Publikationen/Englisch/BOER_recommandation01.pdf

Bioökonomierat (German Bio-Economy Research and Technology Council). (2011). *Bio-Economy Innovation. Bio-Economy Council Report 2010*. Forschungs- und Technologierat Bioökonomie. Available online: http://biooekonomierat.de/fileadmin/Publikationen/Englisch/bioeconomy_council_ report_2010.pdf

Birch, K. (2009). The knowledge—Space dynamic in the UK bioeconomy. *Area*, *41*(3), 273–284. doi:10.1111/j.1475-4762.2008.00864.x

Birch, K. (2012). Knowledge, place, and power: Geographies of value in the bioeconomy. *New Genetics & Society*, *31*(2), 183–201. doi:10.1080/14636778.2012.662051

BMBF (German Federal Ministry for Education and Research). (2010). *Nationale Forschungsstrategie Bioökonomie 2030 (National Research Strategy Bioeconomy 2030)*. Unser Weg zu einer biobasierten Wirtschaft. Available online: https://www.bmbf.de/pub/Nationale_Forschungsstrategie_ Biooekonomie_2030.pdf

BMEL (German Federal Ministry for Food and Agriculture). (2014). *National Policy Strategy on Bioeconomy*. BMEL. Available online: http://www.bmel.de/SharedDocs/Downloads/EN/Publications/ NatPolicyStrategyBioeconomy.pdf?blob=publicationFile

Boehlje, M., & Bröring, S. (2011). The increasing multifunctionality of agricultural raw materials: Three dilemmas for innovation and adoption. *The International Food and Agribusiness Management Review*, *14*, 1–16.

Brown, N., Rappert, B., & Webster, A. (Eds.). (2012). Contested Futures. A Sociology of Prospective Techno-Science. Ashgate.

Capgemini Consulting (CC). (2010). *Roadmap zur Errichtung einer Knowledge-Based Bio-Economy. Nordrhein-Westfalen auf dem Weg in die Umsetzung*. Ministerium für Innovation, Wissenschaft und Forschung des Landes Nordrhein-Westfalen. Available online: http://www.wissenschaft.nrw.de/ fileadmin/Medien/Dokumente/Forschung/Fortschritt/Biooekonomie-Studie.pdf

Chen, H. D., & Gottweis, H. (2013). Stem cell treatments in China: Rethinking the patient role in the global bio-economy. *Bioethics*, *27*(4), 194–207. doi:10.1111/j.1467-8519.2011.01929.x PMID:22092539

Cooke, P. (2007). *Growth Cultures: The Global Bioeconomy and Its Bioregions*. Abingdon, UK: Routledge.

Cooke, P. (2009). The economic geography of knowledge flow hierarchies among internationally networked medical bioclusters: A scientometric analysis. *Tijdschrift voor Economische en Sociale Geografie*, *100*(3), 332–347. doi:10.1111/j.1467-9663.2009.00506.x

Efken, J., Dirksmeyer, W., Kreins, P., & Knecht, M. (2016). Measuring the importance of bioeconomy in Germany: Concept and illustration. *NJAS Wageningen Journal of Life Sciences*, *77*, 9–17. doi:10.1016/j.njas.2016.03.008

European Commission. (2012). *Innovating for Sustainable Growth: A Bioeconomy for Europe; COM 60 final*. Brussels, Belgium: European Commission.

European Commission. (2010). *Innovating for Sustainable Growth: A Bioeconomy for Europe*. European Commission. Available online: http://eur-lex.europa.eu/LexUriServ/LexUriServ.do?uri=COM:2010:2020:FIN:EN:PDF

European Commission (EC). (2010). *Europe 2020. A Strategy for Smart, Sustainable and Inclusive Growth*. COM.

European Council (EC). (2000). *Lisbon Strategy. Presidency Conclusions Lisbon European Council*. Available online: http://www.europarl.europa.eu/summits/lis1_en.htm

Fanuel, M., Khan, B., Singh, N., & Singh, P. (2018). Energy Production in Smart Cities by Utilization of Kinetic Energy of Vehicles over Speed Breaker. *International Journal of Civic Engagement and Social Change*, *5*(2), 1–35. doi:10.4018/IJCESC.2018040101

FORMAS (The Swedish Research Council for Environment, Agricultural Science and Spatial Planning). (2012). *Swedish Research and Innovation. Strategy for a Bio-Based Econom*. FORMAS. Available online: http://www.formas.se/PageFiles/5074/Strategy_Biobased_Ekonomy_hela.pdf

German Presidency (GP). (2007). *En Route to the Knowledge-Based Bio-Economy*. Cologne Paper. Available online https://dechema.de/dechema_media/Cologne_Paper-p-20000945.pdf

Hansen, J. (2014). The Danish biofuel debate: Coupling scientific and politico-economic claims. *Sci. Cult.*, *23*(1), 73–97. doi:10.1080/09505431.2013.808619

Hansen, T., & Winther, L. (2011). Innovation, regional development and relations between high- and low-tech industries. *European Urban and Regional Studies*, *18*(3), 321–339. doi:10.1177/0969776411403990

Hekkert, M., Suurs, R. A. A., Negro, S. O., Kuhlmann, S., & Smits, R. E. H. M. (2007). Functions of innovation systems: A new approach for analysing technological change. *Technological Forecasting and Social Change*, *74*(4), 413–432. doi:10.1016/j.techfore.2006.03.002

Hilgartner, S. (2007). *Making the bioeconomy measurable: Politics of emerging anticipatory machinery*. Academic Press.

Kearnes, M. (2013). Performing synthetic worlds: Situating the bioeconomy. *Science & Public Policy*, *40*(4), 453–465. doi:10.1093cipolct052

Keegan, D., Kretschmer, B., Elbersen, B., & Panoutsou, C. (2013). Cascading use: A systematic approach to biomass beyond the energy sector. *Biofuels, Bioproducts & Biorefining*, *7*(2), 193–206. doi:10.1002/bbb.1351

Khan, B., & Singh, P. (2017). Selecting a Meta-Heuristic Technique for Smart Micro-Grid Optimization Problem: A Comprehensive Analysis. *IEEE Access: Practical Innovations, Open Solutions*, *5*, 13951–13977. doi:10.1109/ACCESS.2017.2728683

Klein, W., Lankhuizen, R. M., & Gilsing, V. (2005). A system failure framework for innovation policy design. *Technovation*, *25*(6), 609–619. doi:10.1016/j.technovation.2003.11.002

Levidow, L., Birch, K., & Papaioannou, T. (2013). Divergent paradigms of European agro-food innovation: The knowledge-based bio-economy (KBBE) as an R & D agenda. *Science, Technology & Human Values*, *38*(1), 94–125. doi:10.1177/0162243912438143

Lösch, A., & Schneider, C. (2016). Transforming power/knowledge apparatuses: The smart grid in the German energy transition. *Innov. Euro. J. Soc. Sci. Res.*, *29*(3), 262–284. doi:10.1080/13511610.2016.1154783

Low, S. A., & Isserman, A. M. (2009). Ethanol and the local economy: Industry trends, location factors, economic impacts, and risks. *Economic Development Quarterly*, *23*(1), 71–88. doi:10.1177/0891242408329485

Marsden, T. (2012). Towards a real sustainable agri-food security and food policy: Beyond the ecological fallacies? *The Political Quarterly*, *83*(1), 139–145. doi:10.1111/j.1467-923X.2012.02242.x

Mathews, J. A. (2009). From the petroeconomy to the bioeconomy: Integrating bioenergy production with agricultural demands. *Biofuels, Bioproducts & Biorefining*, *3*(6), 613–632. doi:10.1002/bbb.181

McCormick, K., & Kautto, N. (2013). The bioeconomy in Europe: An overview. *Sustainability*, *5*(6), 2589–2608. doi:10.3390u5062589

Morrison, M., & Cornips, L. (2012). Exploring the role of dedicated online biotechnology news providers in the innovation economy. *Science, Technology & Human Values*, *37*(3), 262–285. doi:10.1177/0162243911420581

Mumtaz, Z., Bowen, S., & Mumtaz, R. (2012). Meanings of blood, bleeding and blood donations in Pakistan: Implications for national vs. global safe blood supply policies. *Health Policy and Planning*, *27*(2), 147–155. doi:10.1093/heapol/czr016 PMID:21372061

MWK BW (Ministry of Science, Research and Art Baden-Württemberg). (2013). *Bioökonomie im System Aufstellen. Konzept für Eine Baden-Württembergische Forschungsstrategie "Bioökonomie"* [Concept for the Implementation of a Research Strategy on the Bioeconomy in Baden-Württemberg]. MWK BW. Available online: https://mwk.baden-wuerttemberg.de/fileadmin/redaktion/m-mwk/intern/dateien/pdf/ Forschung/ Konzept_Forschungsstrategie_Biooekonomie.pdf

Ollikainen, M. (2014). Forestry in bioeconomy—Smart green growth for the humankind. *Scandinavian Journal of Forest Research*, *29*(4), 360–366. doi:10.1080/02827581.2014.926392

OPST (Office of Science and Technology Policy). (2011). *National Bioeconomy Blueprint: Public Comment*. OPST. Available online: https://obamawhitehouse.archives.gov/administration/ eop/ostp/library/bioeconomy

Organisation for Economic Co-operation and Development (OECD). (2006). The Bioeconomy to 2030.Designing a Policy Agenda. Paris, France: OECD.

Organisation for Economic Co-operation and Development (OECD). (2009). The Bioeconomy to 2030: Designing a Policy Agenda. Paris, France: OECD.

Pollack, A. (2012). White house promotes a bioeconomy. *N. Y. Times*.

Ponte, S. (2009). From fishery to fork: Food safety and sustainability in the 'virtual' knowledge-based bio-economy (KBBE). *Sci. Cult.*, *18*(4), 483–495. doi:10.1080/09505430902873983

Pülzl, H., Kleinschmit, D., & Arts, B. (2012). From a fossil-fuel to a biobased economy: The politics of industrial biotechnology. *Environment and Planning. C, Government & Policy*, *30*(2), 282–296. doi:10.1068/c10209

Rosemann, A. (2014). Standardization as situation-specific achievement: Regulatory diversity and the production of value in intercontinental collaborations in stem cell medicine. *Social Science & Medicine*, *122*, 72–80. doi:10.1016/j.socscimed.2014.10.018 PMID:25441319

Scarlat, N., Dallemand, J.-F., Monforti-Ferrario, F., & Nita, V. (2015). The role of biomass and bioenergy in a future bioeconomy: Policies and facts. *Environmental Development*, *15*, 3–34. doi:10.1016/j.envdev.2015.03.006

Siegmeier, T., & Möller, D. (2013). Mapping research at the intersection of organic farming and bioenergy—A scientometric review. *Renewable & Sustainable Energy Reviews*, *25*, 197–204. doi:10.1016/j.rser.2013.04.025

Singh, P., & Khan, B. (2017). Smart Microgrid Energy Management Using a Novel Artificial Shark Optimization. Complexity. doi:10.1155/2017/2158926

Staffas, L., Gustavsson, M., & McCormick, K. (2013). Strategies and policies for the bioeconomy and bio-based economy: An analysis of official national approaches. *Sustainability*, *5*(6), 2751–2769. doi:10.3390u5062751

Weber, K. M., & Rohracher, H. (2012). Legitimizing research, technology and innovation policies for transformative change – Combining insights from innovation systems and multi-level perspective in a comprehensive 'failures' framework. *Research Policy*, *41*(6), 1037–1047. doi:10.1016/j.respol.2011.10.015

White House. (2012). *National Bioeconomy Blueprint*. Washington, DC: White House.

Wield, D., Hanlin, R., Mittra, J., & Smith, J. (2013). Twenty-first century bioeconomy: Global challenges of biological knowledge for health and agriculture. *Science & Public Policy*, *40*(1), 17–24. doi:10.1093cipolcs116

Winther, T. (2016). *Bioeconomy Strategies and Policies in the Baltic Sea Region Countries, State of Play*. Working Paper No. 1 of the Baltic Sea Regional Bioeconomy Council. Available online: http://bsrbioeconomy.net/resources/2016_docs/Working_Paper_1_%20BSR_Council.pdf

Zilberman, D., Kim, E., Kirschner, S., Kaplan, S., & Reeves, J. (2014). Technology and the future bioeconomy. *Agricultural Economics*, *44*(s1), 95–102. doi:10.1111/agec.12054

KEY TERMS AND DEFINITIONS

Bio-Ecology: It is the stream of biology which studies the connection between various living organisms and their environment.

Bio-Refinery: It represents all economic actions performed from research and scientific actions focused on biotechnology.

Bio-Resources: These are biogenic resources other than fossil fuels those can be utilized for various purposes.

Bio-Technology: The utilization of biological processes for commercial and industrial purposes.

Chapter 2
Strategies of Sustainable Bioeconomy in the Industry 4.0 Framework for Inclusive and Social Prosperity

Erhan İşcan
https://orcid.org/0000-0001-6068-6698
Cukurova University, Turkey

ABSTRACT

The last two decades have witnessed unpredictable transformations in knowledge-related areas. The main reason for this change is the fourth industrial revolution, a knowledge revolution affecting fields like the bioeconomy. The third industrial revolution, which induced the use of fossil-based energy sources, created a major global problem. Likewise, the third industrial revolution introduced the problem of the excess usage of food, animals, water, and other resources. Industry 4.0 offers an efficient solution to excessive tendencies. This chapter aims to analyze changes and offer strategies in the bioeconomy framework within Industry 4.0.

DOI: 10.4018/978-1-5225-7958-8.ch002

INTRODUCTION

Food is a basic human need. Therefore, the agricultural sector has become the most important part of the economy. The share of agricultural production in the economy has been declining over the last two decades. However, investments have boosted productivity. This growth is negatively affected by the degradation of natural resources, global climate change, excess use of pesticides, and loss of biodiversity. Additionally, technological progress has impacted human needs, increased productivity, and changed production methods. Technological progress has changed the path of the agricultural sector by affecting the production methods.

The Food and Agriculture Organization of the United Nation reported that the human population is expected to increase to 10 billion by 2050 in the "The Future of Food and Agriculture – Alternative Pathways to 2050" report (FAO, 2018). Food security will be the main problem faced by governments. Most of the literature estimates more severe crises in both numbers and intensity, giving rise to poverty and hunger throughout the world. In this context, it becomes very important to analyze the importance and impact of technology on agricultural production, agricultural productivity, and the share of the agricultural sector in the economy.

This study analyzes the impact of technology on the agricultural sector. It highlights sustainable development and discusses a sustainable bioeconomy. This study offers technology-based policies for a sustainable bioeconomy with inclusive growth.

BACKGROUND: A HISTORY OF REVOLUTION, EMERGENCE OF INDUSTRY 4.0, AND THE AGRICULTURAL SECTOR

History reveals seven critical transitions of humankind. Three of these transitions focus on agriculture and the volume of food production. Four of the transitions are industrial, focusing on mechanization. First, the Neolithic revolution (or the first agricultural revolution) is a basic and critical revolution that changed humankind from hunter-gatherers to a farming society. About 14,000 years ago, humankind domesticated plants like wheat, barley, peas, and livestock. The Neolithic revolution increased life expectancy and improved human life by establishing cities with more food. These developments promoted new work, jobs, merchandising, and trade. Specializations introduced skilled professions. Society became more complex, requiring new rules. This need brought about an authoritative government.

After 15,000 years, humankind witnessed the second agricultural revolution, which coincided with the industrial revolution. Between the 17th century and the green revolution of the 1940s, farmers became more capable. They increased productivity with ideas like selective breeding, crop rotation, and fertilizers. The industrial revolution also brought farm machines and tools, increasing productivity and production in a short amount of time. Demographics changed as farmers replaced machines with workers. More food affected the rapid growth of cities as people stopped producing food for self-consumption. These changes in society and production promoted the industrial revolution.

The third revolution, termed the green revolution, affected the life of humankind. Genetically engineered crops and/or genetically modified organisms opened a new stage in agriculture and the economic sector. The yield of wheat increased fourfold in 25 years; many countries became self-sufficient. The green revolution satisfied an increasing demand of food due to the increasing population of a healthier society.

The three agricultural revolutions increased the quantity of food and promoted the industrial revolution, the key driver of modern society's way of life. In the middle of the 18th century, the economic system changed radically with the invention of James Watt's steam machine. The system moved from an agrarian economy to an industrial economy. New tools and/or machines enabled large-scale production using new materials. Factories grew as labor became more important. Transportation, communication, and steam-powered machines and vehicles were also key to the first industrial revolution. The growth of agricultural and industrial production was a new source of wealth. Increasing productivity promoted international trade and national prosperity.

The second industrial revolution was a technological shift during the 19th century. Electricity changed the world through many inventions between 1870 and 1910. Mass production, chemical synthesis, synthetic raw materials, oil as a substitute and primary energy source, long-distance communication, the first flight, Edison's lightbulb with tungsten-based filament, and Ford's Model T assembly line are examples that reflect the characteristics of the second industrial revolution.

The 20th century witnessed the third industrial revolution with the invention and use of nuclear power, the Internet, automated machines, three-dimensional (3-D) printers, and affordable computer equipment and electronics. Digitalization and digitalized automation transformed the manufacturing process. Fitzsimmons (1994) stated that a combined power of information technology fueled the third industrial revolution. Technology and information providers empowered people to find, retrieve, share, and use data in ways that enriched their lives (Fitzsimmons, 1994). This information-based economic structure was termed the "digital economy."

The last two decades witnessed the emergence of the fourth industrial revolution's radical change in technology. The beginning of the 21st century brought smart factories with cyber-physical systems and the Internet of things (IoT). This revolution is termed the "fourth industrial revolution." According to Schwab (2016), this revolution differs from the other ones because it combines physical, digital, and biological worlds. Advanced robotics, artificial intelligence (AI), synthetic biology, "smart" items, and a sharing economy are elements of the new era.

The Organisation for Economic Cooperation and Development emphasized the importance of space economy for the future (OECD, 2016). AI, in particular, can play a vital role in shaping the future. Every sector in the economy has been affected by technological changes. These sectors are causing shifts in the world economy.

Likewise, technology changed the weight and structure of the economy's agricultural sector. However, these changes were influenced later than the other sectors. In the beginning of the fourth industrial revolution, economic agents of the agricultural sector did not internalize new technology. In turn, they could not apply to agricultural production processes. During the last decade, the structure of the agricultural sector absorbed and utilized new technologies. This change in the use of technology also changed the agricultural sector to an agro-food sector, giving rise to the bioeconomy. This new kind of agricultural sector crossed borders via corporate sectors and internationalization. It became the key understanding of the new technology-driven agricultural sector. Additionally, the share of agriculture increased in foreign trade through the sector's change. Technological changes in the agricultural production processes affect agriculture and food policies by increasing production, efficiency, security, and competitiveness. An understanding of these policies is required to absorb the changes.

Johnston and Mellor (1961) noted an interrelationship between agricultural and industrial development. Additionally, they found that agriculture's role in the process of economic growth must be analyzed (Johnston & Mellor, 1961). Since ancient times, the agricultural sector has evolved within industrial revolutions. Every industrial revolution, like every sector in the economy, has advanced the agriculture and food sector. Conventional techniques have transformed to modern technological processes in the last two centuries. After 70 years, we still follow the same statement. However, we add the word "fourth" before "industrial revolution." In this framework, it is important to analyze the effects of technological changes on the agricultural sector.

The production of goods and services is impacted by scientific improvements after every industrial revolution. Inventions in chemistry, biology, and engineering have led to a new science field termed "biotechnology." Agriculture is utilized in every type of invention throughout the history of science. For example, inventions like the microscope and rubber tire somehow increased agricultural production. After the third revolution, automated farm equipment, microchip-built hydraulic equipment, and automated feeding systems promoted the crop industry, dairy industry, and livestock industry. During the third revolution, the agricultural industry evolved into an agro-food industry. Biotechnology is the prominent element that advanced agriculture after the third industrial revolution. The fourth industrial revolution changed the definition of domestication as it turned to the deliberate control of biological systems for producing efficient, disease-resistant, and cleaner agricultural food and services. Biotechnology affected the traditional industries of food processing and fermentation.

In the first and second revolutions, scientists worked for the science of agriculture. The third revolution changed the number of corporations, institutions, universities, foundations, and enterprises that researched in biotechnology studies. After the early 1960s, many institutions and corporations began to research biotechnology. Today, over 12,000 firms conduct research in OECD countries (OECD, 2018).

The fourth industrial revolution provided research and definitions in the literature. The fourth revolution, or Industry 4.0, introduces smart systems to monitor processes through cyber-physical systems. In other words, physical systems communicate and cooperate with each other and with humans via the Internet and cloud. More than 100 billion connected devices will transform human life, work, communication, needs, and thoughts. Industry 4.0 is different because it is more complicated than previous revolutions. Industry 4.0 improves human life through biology, technology, and industrial automation. The key of Industry 4.0 is its emergence of AI to drive cars through autonomous controllers, cure humans with nanobots, provide self-connected hardware, etc. The extensive list of improvements is a key difference of Industry 4.0. The first three revolutions changed the world with several innovations like steam machines and electricity. Today, innovations are anywhere and at any time across the globe. Industry 4.0 affects work models by establishing relations between equipment and environments. Cyber-physical systems can decentralize decision-making through self-ruling systems. Therefore, interoperability and IoT are important in the fourth revolution. This new perspective induced a quest for clean energy and nanotechnology. New kinds of production are based on clean energy as we enhance materials with nanotechnology.

EFFECTS OF INDUSTRY 4.0 ON THE BIOECONOMY AND IMPORTANCE FOR INCLUSIVE, SUSTAINABLE GROWTH

The fourth industrial revolution has had complicated effects on the distribution of production, the amount of produced goods and services, and the quantity of sources used for production. Industry 4.0 brought new problems to humankind as it offered solutions to both existing and new problems. While the demand for sources for production increased, biotechnology solved the problem of scarce sources by increasing efficiency. On the other hand, biotechnology will solve problems caused by the industrial revolutions, including a reduction in fuel use, carbon dioxide emissions, compaction of soil, and erosion.

Industry 4.0 may cure current and future problems throughout the world. Therefore, understanding the effect of the Industry 4.0 bioeconomy is very important for inclusive and sustainable growth. Renewable resources will take the place of fossil-based or conventional resources. This will trigger a new global economic growth, with prepared nations receiving the largest share. In 2050, with a world population of 10 billion people, the demand of sources, particularly biotechnological sources, will increase. The goal of bioeconomy policies is simple: Have renewable resources, biotechnology, and biobased products by 2050.

According to OECD, the bioeconomy involves three elements: (1) biotechnological knowledge; (2) renewable biomass; and (3) integration across applications (OECD, 2009). These elements are clearly involved with the Industry 4.0 framework. In the same report, OECD stated that biotechnology can support sustainable development by improving environmental efficiency. Examples of improvements include the removal of toxic compounds, reductions to soil erosion, the use of genetic fingerprinting to manage and protect wild fish stocks, and biotechnological applications to reduce greenhouse gas emissions (OECD, 2009). These ideas clearly summarize the valuable effects of Industry 4.0 on the bioeconomy. The United Nations Industrial Development Organization (UNIDO), in their report on Industry 4.0 and energy, stated that:

Industry 4.0 could open new opportunities for encouraging the deployment of renewable energies, e.g. when companies harmonize their production cycles with peak generation times to better match demand for power with the actual supply. In that manner, Industry 4.0 can help cope with the significant fluctuation of energy supplied by renewable energy sources—one of its big challenges. (UNIDO, 2017, p. 43)

The European Union (EU) established a knowledge-based bioeconomy definition to reflect the European bioeconomy. This definition includes the new knowledge-dependent environment. Europa highlighted the science, technology, research and development, and high-technology industry to support innovation, progression in food technologies and products, and animal breeding technologies. It also focused on establishing a strong chemical and manufacturing industry base (EU, 2018).

Biotechnological Knowledge and Industry 4.0

A sustainable bioeconomy based on biotechnological knowledge will result in the development of new products. Sustainable development requires a broader definition of the bioeconomy and a wider understanding of its necessary knowledge base (Schmidt, Padel, & Levidow, 2012). Industry 4.0 affects the development of knowledge regarding biotechnology. Production and circulation of knowledge has increased through Industry 4.0 due to intensive research, development, and innovation.

On the other hand, increased innovative ecosystems have impacted the creation of biotechnological knowledge. If the availability of food will be a problem by 2030, important issues to address include improved use of efficient resources, improved yield, reliable harvests for increased resilience, improved plant health and pest resistance, and reduced environmental footprints (Malyska & Jacobi, 2017). These issues are main topics related to the bioeconomy and Industry 4.0.

Industry 4.0 changed how we produce information and technology in many sectors of the economy. Industry 4.0 introduced scientific trends, technological changes, and innovative approaches with an unlimited circulation of knowledge.

It is important to link ecology, industry, and society with resource-efficient and technological processes. The bioeconomy's potential becomes the most important component for a sustainable future. New technologies and processes enable the bioeconomy's potential to provide sustainable economic growth. The bioeconomy is seen as the best possible use of materials and the flow of these materials.

Industry 4.0 will provide a base for a sustainable bioeconomy by enabling products through new knowledge-based technologies that ensure inclusive and social prosperity (Schütte, 2018). Schütte termed this process "biologization" in his seminal paper, noting that "… biologization will be as significant as a cross-cutting approach as digitalization already is today" (Schütte, 2018, p. 84). For sustainable inclusive and social prosperity, new knowledge related to the bioeconomy is crucial. Industry 4.0 induced digitalization, the fundamental component of the bioeconomy.

Renewable Biomass and Industry 4.0

Production has changed from a fossil- to biobased focus. This change is the second element of the bioeconomy. In the last century, urbanization increased with the world's population. In turn, the third industrial revolution induced energy usage. These advances affect resources, energy and water use, waste generation, and the demand of agricultural products. The excess use of resources and climate changes brought attention to the renewable biomass as policymakers reviewed earth's future.

Industry 4.0, opposite to the third industrial revolution, induced energy efficiency and production of renewable resources. The issue of renewable biomass is essential for developing countries for two reasons. First, developing economies based on the agricultural sector produce more biomass. Second, producing renewable biomass requires biotechnology that is dependent on research and development. This reason is more important for developing countries because renewable biomass will induce research and development on high technologies. These efforts will change the environment of a national economy.

Industry 4.0 makes it easier for developing countries to attain knowledge. Renewable biomass is powerful in integrating an economy, innovation, agricultural sectors, and high technologies. Moreover, developing countries will decrease their dependency on petroleum as a major resource. The use of appropriate technologies in processing and manufacturing will improve efficiency of production and quality of products, as well as facilitate trade and international development cooperation (Lokko et al., 2018). Developed countries, which are already aware of problems related to future food and resource security, are initiating a knowledge-base to establish technological biobased sectors.

The EU's bioeconomy definition shows the effect and importance of Industry 4.0. The knowledge-based bioeconomy is defined as "life sciences and biotechnology knowledge converging with other technologies to transform into new, sustainable, eco-efficient, and competitive products" (EU, 2018). Therefore, the bioeconomy is based on transformation of knowledge to new sustainable products. Producing renewable biomass is an essential element for sustainability as it promotes inclusive and social prosperity. In the agriculture and agribusiness sectors, biotechnology applications play a significant role by increasing productivity and diversification of agriculture produce, as well as reducing their environmental impacts (Lokko et al., 2018).

Antonio Guterres, Secretary General of United Nations, remarked in the World Economic Forum that "... the best prevention for conflict and the best prevention for other negative impacts on societies is, of course, sustainable and inclusive development." (UN, 2016) Renewable biomass and Industry 4.0 share a foundation based on technological innovation. Digital technologies offer new production techniques, higher collaboration and investment in research, innovation, and skills for an advanced knowledge base. Policies connecting Industry 4.0 and renewable biomass will stimulate carbon-reducing industries through digitalization, innovation, networks, knowledge sharing, and research.

Integration Across Applications Related to Industry 4.0

This section discusses the integration of knowledge and applications based on generic knowledge and value-added chains across applications. Primary production, health applications, and industrial applications are three fields related to the digitalization of Industry 4.0 (OECD, 2009). Biotechnology applications are well-known sides of platform technologies. OECD indicated that these platform technologies are main research tools and techniques for modern biotechnology and the future of the bioeconomy. Support for integration requires coordinated actions that draw on the expertise of numerous government ministries, including agriculture, education, environment, health, industry, natural resources, and research (OECD, 2009).

Digitalization, AI, and cyber-physical systems are key drivers of this new era. They will also be the main promoters for the bioeconomy in the future. Using several techniques, biotechnology will be the most important factor for fighting against food shortages. Biotechnology is the main area affected by Industry 4.0. This digital revolution is a major element impacting the bioeconomy through new technologies. The main fields studied by biotechnology include the production of new crop varieties, development of new forestry and tree varieties, diagnosis of plants to decrease economic damage, and animal breeding, propagation, and diagnosis.

On the other hand, therapeutics, diagnostics, pharmacogenetics, functional foods, and medical devices are the main fields impacted by biotechnology. These are strongly linked with each other through integrated platforms. Without the contribution of digitalization, there will be limited chances to solve future problems of humanity through the bioeconomy. Industry 4.0 will be an important basis for the bioeconomy by providing solutions, approaches, and platforms. Industry 4.0 will continue to change the level of technology with paradigms; these evolutions will be the key for the bioeconomy.

Schwab (2017) stated that Industry 4.0 includes smart, connected machines and systems, as well as the fusion of technologies across physical, digital, and biological worlds (Schwab, 2017). New changes, technologies, and paradigms with Industry 4.0 will continue to shift in relation to the bioeconomy. Its new framework will integrates all agents related to the bioeconomy. For instance, 3D printing and advanced robotics are a physical megatrend of Industry 4.0. 3D printing may be a new source of human organs. Big data and robots, particularly AI, may be a new source of biotechnology. New interactions between humans and machines present a potential shift in the bioeconomy. New megatrends may be created as innovations in the physical world shift to digitalization and biological spheres. These progressions and shifts will continue to impact the physical world. This interrelationship is essential for the current and future bioeconomy.

Human Health and the Bioeconomy

The third industrial revolution challenged the world by using electronics. Biotechnology emerged by studying recombinant DNA with the power of electronics. In the early 1950s, scientist studied human DNA. This can be accepted as a starting point in biotechnology. In the 1960s and 1970s, several scientific improvements emerged surrounding DNA. These improvements rocketed the evolution of the pharmaceutical industry. As a result, many pharmaceuticals were invented in the last quarter of the 20th century.

According to OECD (2009), five areas apply to biotechnology. The first area, therapeutics, consists of three groups: (1) biopharmaceuticals; (2) experimental treatment; and (3) small-molecule therapeutics. The second area is diagnostics. A diagnostic test of modern biotechnology identifies genetic or nongenetic diseases. The third, pharmacogenetics, examines the interaction between genes and drugs. The fourth, functional foods and nutraceuticals, include fish oil and functional foods with added nutrients. The fifth, medical devices, include surgical and diagnostic instruments (OECD, 2009). These areas are very important for human development and technological improvements.

Industry 4.0 also introduced disruptive technologies to human life. These include regenerative medicines like tissue engineering, stem cell treatment, and gene therapies. Industry 4.0 is a main promoter of these experimental technologies. A continued supply of these new applications will be the outcome of the interaction. The effects of Industry 4.0 biotechnology will highly important to economic sectors (OECD, 2009).

STRATEGIES OF A SUSTAINABLE BIOECONOMY WITHIN THE INDUSTRY 4.0 FRAMEWORK

OECD (2018) stated that the bioeconomy concept emerged from a niche interest. It evolved into the political mainstream with over 50 countries publishing bioeconomy policies and intentions (OECD, 2018). Per this statement, every nation across the globe needs a policy on the bioeconomy. The same report highlighted the biotechnology-centric vision of an economic activity that spreads across several key sectors and policy families (OECD, 2018).

The bioeconomy policy framework must be developed within the changing environment. A sustainable bioeconomy became essential for sustainable growth promoting inclusive and social prosperity. With an increasing population, climate change, food security, and energy security create environmental difficulties for policymakers. In this context, knowledge-based innovative bioeconomy policies must include a wide range of goals and conclusions.

Creating and Promoting the Knowledge-Based Sustainable Bioeconomy Vision

The last two decades have witnessed the emergence of a knowledge-based economy. Knowledge-based economics is based on the use of knowledge and information on production. High-technology investments, production, industries, and labor, as well as knowledge distribution and information society, are keywords related to knowledge-based economics and Industry 4.0. The knowledge-based bioeconomy shares the same basis as the knowledge-based economy. The knowledge-based bioeconomy will contribute to food production for an increasing population. It will fight climate change through the power of high-technology and knowledge.

In this framework, creating a vision driven by a knowledge-based sustainable bioeconomy becomes more essential. If the bioeconomy focuses on production and use of renewable sources in all economic sectors, it must be based on knowledge like the knowledge-based economy. Being knowledge-based is essential to feed the entire world, fight climate change, use resources efficiently, and/or protect biodiversity. For the future, the only way of producing more to increase prosperity is via the bioeconomy.

The vision to contain and accept the knowledge-based bioeconomy will increase economic output with new materials, products, and integrated value-adding networks.

Much of a national bioeconomy strategy depends on this knowledge-based mentality. According to OECD (2009), most nations emphasized the role and effects of access to technological knowledge in agriculture, health, and/or industry. OECD (2009) indicated that basic and applied research with a multipurpose infrastructure and education will generate an inclusive and social prosperity.

After this decade, every nation will need to implement the knowledge-based bioeconomy in their growth and development strategies. Growth strategies increase economic output and development through shared and increased wealth. Increased data volume, computing power, and integration are essential to the growth and development of an economy. Therefore, the bioeconomy needs a knowledge-based vision.

On the other hand, Industry 4.0 identified a difference in the disciplines: integration and knowledge sharing. These are important because innovation in many areas has the potential to be interrelated. Accordingly, strategies, goals, and guiding principles must be created with a broad vision for a sustainable bioeconomy. Additionally, this vision must include creativity and innovation for improvements. These should be shared as key drivers of both the bioeconomy and economy.

Increasing the Capacity of Innovation for a Sustainable Bioeconomy

Innovation is one the most discussed components of human life. This article discusses innovation based on its shaping of the previous century. Development and sustainability of the bioeconomy depend on innovation and its capacity. Biotechnology needs a high amount of research and development to produce goods and services with new patents.

A sustainable future is only possible by reconciling ecology and economy for secure food and resources. To increase the capacity of innovation, we must promote research and development with an established and suitable environment. Establishing an innovation policy is an important component to a successful and sustainable bioeconomy. Through innovation, the traditional fossil-based economy will increase its demand on renewable sources. This demand will be satisfied without endangering food security or the natural basis for life (Schütte, 2017).

The European Commission (EC) set research and development as a primary content of the strategy and action plan for the sustainable bioeconomy. At the same time, the EC stated that a large knowledge base with the ability to innovate will support the success of economic growth and development. Today's economy still depends on basic natural biological resources and/or traditional raw materials. However, new technologies, including life sciences and biotechnology, are transforming the economy. This is important because only innovation will lead to economic growth or development. If a nation lacks these two components, there will be no prosperity in the future. Therefore, supporting research and development for innovation became the primary component of the EU's strategy and action plan to promoting a sustainable bioeconomy.

The EU stated that Europe plays a pioneer role in research and development. However, there is a problem when changing to high value-added commercial products. The development of a bioeconomy is extremely dependent on research and development. Increases in the spending and investments on biotechnology will promote the economic performance of a nation. Through Industry 4.0, the bioeconomy will become smart. Its development will depend on excellent innovation and research induced by science, industry, and society (Aguilar, Wohlgemuth, & Twardowski, 2018).

The bioeconomy cannot develop without science and technology's impact on innovation. The future of the bioeconomy requires shared knowledge to produce new knowledge and innovative biobased products. Policymakers must create new policies to promote capacity of innovation. This is the intersection of Industry 4.0 and the bioeconomy. Aguilar et al. (2018, p. 182) noted, "Bioeconomy requires a subtle equilibrium between science push, supported by the public sector, and market and social pull, stimulated by industry and investors as well as by policy makers and civil society."

Creating and Promoting Governance and Participation Between Agents While Changing Conventional Policies

OECD (2018) defined governance as the system of regulations and policies that influence the development of the bioeconomy. They linked its success to governance and economic competitiveness of biotechnological innovations (OECD, 2018). Aguilar et al. (2018) emphasized the importance of governance, noting that policy decisions affect the bioeconomy. Like any other political decision, the support of governance is required in the respective country and/or regions (Aguilar et al., 2018).

Agents of the knowledge-based bioeconomy are scientists, government, the private and public sectors, citizens, and other nations. Defining agents is essential so governance will be understandable and easy. There must be cross-sector cooperation and collaboration between agents to achieve long-term development. The public and private sectors should be encouraged to build relationships to increase research and development, which will contribute to biotechnology.

Industry 4.0 is very different from the other industrial revolutions due to the important role played by governance. It is not surprising, therefore, that many policies in the new economy paradigm require the active participation of citizens.

OECD indicated the need to create an active and sustained dialogue between industries on socioeconomic and ethical implications, benefits, and requirements of biotechnologies (OECD, 2009). The EU accepted the same framework, emphasizing the collaboration between actors of bioeconomy policies. Likewise, establishing coherence between policies at national and global levels is a major point of governance. On the other hand, there will be many conflicts between agents. There will also be important decisions regarding strategies and trade-offs in the design and development of sustainable bioeconomy policies. Therefore, creating governance became key in building coherent and integrated policies dependent on national and global collaboration.

The EU stated that a responsible bioeconomy calls for participatory models that engage citizens and end users to reinforce relationships between the scientific society and policymakers (EC, 2012). Information flow between agents will impact system sustainability. Hence, it is essential for consistency and coherence of strategies and policies. Engagement between manufacturing sectors and the political impetus will promote the benefits of a bioeconomy.

There will be no development of the bioeconomy without acceptation, understanding, and promotion of agents. Reducing fossil-based resource use, accepting biotechnological improvements (for cloning and genetic modification), adopting technological improvements to breed plants, and demanding zero-fossil-based products is only possible with agents who are aware. Lainez, González, Aguilar, and Vela (2018), when evaluating the Spanish bioeconomy strategy, stated that:

society must be given the opportunity to observe its development during the implementation process. This would allow consumers to be more informed regarding the new range of products which will gradually become available and appear on the market. (p. 92)

For instance, policymakers must clearly inform citizens about improvements in health and biotechnology because a society that is uninformed about issues like genetic modification, cloning animals, and pharmacogenetics will strictly deny these developments. This will slow the development of the bioeconomy.

Gustafsson, Stoor, and Tsvetkova (2011), when studying the credibility of the bioeconomy, argued that:

credibility of the bio-economy needs to be increased. At the moment the bioeconomy suffers from a lack of credibility. There is also very little understanding of the bio-economy and its potential. This influences legislation and decisions on the municipal level. (pp. 61-62)

They offered a simple solution to this problem, stating that "governmental support in the form of solid policies, legislation, recommendations, and visions regarding the bio-economy on the country level would contribute significantly to the concept's credibility" (Gustafsson et al., 2011, pp. 61-62). Evidence in the literature supports the idea that governance and participation is vital to the bioeconomy.

Establishing Biobased Technology Transfer Offices

OECD stated barriers to biotechnology innovation can be reduced by identifying factors that prevent the development of innovative markets. This is achieved by evaluating possible policy actions that support access to knowledge. In addition, institutions can be encouraged to adopt intellectual property guidelines for rapid innovation and the sharing of knowledge.

OECD (2009) noted the importance of knowledge spillover across research disciplines and commercial applications. This spillover integrates biotechnology and research. Innovation processes and research on the knowledge base must be available for sharing. Van Lancker, Wauters, and Van Huylenbroeck (2016) stated that openness provides many benefits to agents by increasing the cost of research and development, increasing the identification of potentially valuable opportunities for innovation, providing access to knowledge, and facilitating integration and cooperation between organizations. More benefits can be identified regarding collaboration, integration, and sharing networks. However, only an institution can create, organize, and increase the expected benefits. This institution is a biobased technology transfer office.

After Industry 4.0 changed the economic environment, knowledge-based economics required different types of institutions. The bioeconomy technology transfer offices are new institutions to frame and organize the bioeconomy. This office analyzes the environment, determines problems, sources, and needs, and integrates agents through supervision. In doing so, the institute will energize the region's potential.

The offices are similar to technology transfer offices. These offices are organizations in which activities related to the commercialization of biotechnology academic research results are carried out efficiently and rapidly. It can be found among universities, research centers, the private sector, researchers and entrepreneurs, investors, and industrialists. These offices enable manufacturing. Investors can meet with researchers to transfer biobased industrial knowledge. Biobased technology transfer includes a series of interactions (i.e., knowledge, experience, and equipment flow) among stakeholders, including the public, firms, financial institutions, research and education institutions, and nongovernmental organizations.

The commercialization of scientific studies in the bioeconomy is difficult compared to other areas. For this reason, biobased technology transfer offices ensure that technology in research institutions is transferred to the industry. Arujanan and Singaram (2018) argued that the biotechnology sector is naturally multilayered, multidisciplinary, and complex, making it difficult to unlock its true potential and provide quality services in a short span of time. This is a challenge for the bioeconomy in Malaysia. Additionally, there is a mismatch between industry needs and research priorities within universities (Arujanan & Singaram, 2018). These specific offices have the potential to solve this kind of problem.

FUTURE RESEARCH DIRECTIONS

Industry 4.0 and Web 2.0 are relevant topics. The future is changing regarding technologies and innovations. In the long term, the bioeconomy will be affected by technological developments, diffusion of knowledge sharing networks, business models, market structures, and regulations. Improvements in biotechnology will be important due to the rising demand and increased agronomic stresses from climate change. OECD noted a long-term increase in the cost of fossil fuels due to a decline in the supply of low-cost sources of petroleum, an increase in demand for energy, and restrictions on the production of greenhouse gases. These could create a growing market for biomass, including nonfood crops like grasses and trees as feedstock for biofuels, chemicals, and plastics (OECD, 2009).

Improvements, changes, and problems will evolve within agricultural technologies. They will also increase investments in the bioeconomy. These investments will focus on high-technology agricultural production by biotechnology through genetic modification. It will aim to increase crops, clone animals, and produce zero-carbon level energy sources.

Combining biotechnology with other technological fields will be essential for the future. The next decade will introduce a shift in paradigms and a change in practice. These shifts will create a broader approach using high-tech bioeconomy. The EU indicated that "growing knowledge, bioinformatics, and the strong interaction of engineering sciences with life sciences will open avenues to new products and applications" (EC, 2012).

In this framework, it is essential to increase the share of multidisciplinary and cross-sectoral research and innovation to address the complexity and interconnectedness of societal challenges. This will improve the existing knowledge-base and develop new technologies (EC, 2012). Future strategies include the building of human capacity to support growth, integration of bioeconomy sectors, promotion of uptake, and diffusion of innovation in bioeconomy sectors (EC, 2012). EC has 10 priorities for the future of Europa. The bioeconomy is found in three of the 10 priorities.

1. **Jobs, Growth, and Investment:** The innovative bioeconomy boosts jobs and prosperity. There will be a potential for biobased markets and products, including renewable biomass.
2. **Energy and Climate:** This union makes energy more secure, affordable, and sustainable. Using coordinated research, Europe must diversify its sources of energy and support breakthroughs in low-carbon technologies. Replacing fossil raw materials with biological resources is an indispensable component of a forward-looking climate change policy.
3. **Economics and Money:** More fair economic and monetary unions promote stability of the economic environment. Innovative biobased food industries raise the GDP from 16% to 20%. It creates a circular, resource-efficient economy. The food and drink industry is the largest manufacturing sector in the EU (EC, 2018).

The bioeconomy is cross-sectoral and multidisciplined. It surrounds global issues as it changes the world to a smart-blue economy. To understand the future, interaction between Industry 4.0 and the bioeconomy must be fully analyzed.

Sachsenmeier (2016), in his paper about Industry 5.0, stated that "Industry 5.0 discussions touch on the very essence of humanity's existence, physical integrity, and relationship with nature. This debate seems theoretical, yet it will soon come to the fore" (Sachsenmeier, 2016, p. 229). The future direction is quite complex. Therefore, we must start to think about it today.

CONCLUSION

This chapter highlights the effect of Industry 4.0 on the bioeconomy. It also analyzes changes in the bioeconomy paradigm. Economic, social, and the physical environment changed with Industry 4.0 as they evolved into very different stages. Industry 3.0 induced the usage of fossil-based energy sources, which became a major global problem. Likewise, Industry 3.0 brought us the problem of excess usage of food, animal, water, and other resources. Industry 4.0 brought a solution to our excess behavior, encouraging us to be more efficient.

Two decades ago, parallel with Industry 4.0, the bioeconomy entered the literature. The major reason behind its emergence was the fear of finite amounts of fossil fuels, climate change, and scarcity of resources. The UN, EC, OECD, and over 50 nations prepared bioeconomy policy documents, reports, and strategies. The bioeconomy became an emerging topic in academic literature. Researchers contributed scientific studies emphasizing smart, sustainable, and inclusive growth. Today, the bioeconomy is a major and adjoining topic. It offers a new perspective to policymakers.

The development of the bioeconomy is an important topic for policymakers in ensuring sustainable development. The world is shifting from fossil-based nonrenewable resources to biobased renewable resources. This change requires strategies and policies. The bioeconomy has the potential to increase jobs, growth, and prosperity by creating functional strategies and/or policies. This chapter proposes four strategies for policymakers.

1. **Creating and Promoting the Knowledge-Based Sustainable Bioeconomy:** This vision emphasizes the importance and content of a bioeconomy vision. Accordingly, strategies, principles, goals, and guiding principles must be created with a broad vision for a sustainable bioeconomy. This vision must include creativity and innovation for improvements. These must be shared as key drivers of the bioeconomy and economy.
2. **Increasing the Capacity of Innovation for a Sustainable Bioeconomy:** This strategy discusses the effects of innovation on the bioeconomy. The development of the bioeconomy is extremely dependent on research and development as it increases spending and investment on biotechnology to promote national economic performance.
3. **Creating Governance Between Agents While Changing Conventional Policies:** This strategy emphasizes that problem of information flow between agents. This problem will prevent the system from attaining sustainability. Hence, consistency and coherence of the strategies and policies is essential. Engagement between manufacturing sectors and the political impetus will promote the benefits of the bioeconomy.

4. **Establishing the Bioeconomy Technology Office:** There are many benefits, including collaboration, integration, and sharing networks. However, only an institution can create, organize, and increase the expected benefits. This institution is the bioeconomy technology office. Industry 4.0 changed the economic environment. Accordingly, knowledge-based economics need a different type of institutions. The bioeconomy technology office, which is a new type of institution, will frame and organize the bioeconomy.

It is clear that the bioeconomy will be a major area for nations as they study lessons, prepare strategies, and create policies. The authors understood that the bioeconomy is advancing through Industry 4.0. However, it is estimated that they will interact. The nation that created policies will have smart, inclusive, sustainable, and social prosperity.

ACKNOWLEDGMENT

This research was supported by the Research Fund of Cukurova University [Project Number: SBI-2019-11685].

REFERENCES

Aguilar, A., Wohlgemuth, R., & Twardowski, T. (2018). Perspectives on bioeconomy. *New Biotechnology, 40*(Pt A), 181.

Arujanan, M., & Singaram, M. (2018). The biotechnology and bioeconomy landscape in Malaysia. *New Biotechnology, 40*(Pt A), 52-59.

European Commission. (2012). *Innovating for Sustainable Growth - A Bioeconomy for Europe*. Luxembourg: Publications Office of the European Union.

European Union. (2018). *Bioeconomy Policy*. Retrieved from https://ec.europa.eu/research/bioeconomy/index.cfm?pg=policy

Fitzsimmons, J. (1994). Information technology and the third industrial revolution. *The Electronic Library*, *12*(5), 295–297. doi:10.1108/eb045307

Gustafsson, M., Stoor, R., & Tsvetkova, A. (2011). Sustainable bio-economy: potential, challenges and opportunities in Finland. *SITRA Studies, 51*, 64.

Johnston, B. F., & Mellor, J. W. (1961). The role of agriculture in economic development. *The American Economic Review*, *51*(4), 566–593.

Lainez, M., González, J. M., Aguilar, A., & Vela, C. (2017). Spanish strategy on bioeconomy: Towards a knowledge based sustainable innovation. *New Biotechnology*. PMID:28552816

Lokko, Y., Heijde, M., Schebesta, K., Scholtès, P., Van Montagu, M., & Giacca, M. (2017). Biotechnology and the bioeconomy—Towards inclusive and sustainable industrial development. *New Biotechnology*. PMID:28663120

Małyska, A., & Jacobi, J. (2017). Plant breeding as the cornerstone of a sustainable bioeconomy. *New Biotechnology*. PMID:28690155

OECD. (2009). *The Bioeconomy to 2030: Designing a Policy Agenda*. Paris: OECD Publishing; doi:10.1787/9789264056886-

OECD. (2018). *Meeting Policy Challenges for a Sustainable Bioeconomy*. Paris: OECD Publishing; doi:10.1787/9789264292345-

Sachsenmeier, P. (2016). Industry 5.0—The Relevance and Implications of Bionics and Synthetic Biology. *Engineering*, *2*(2), 225–229. doi:10.1016/J.ENG.2016.02.015

Schmidt, O., Padel, S., & Levidow, L. (2012). The bio-economy concept and knowledge base in a public goods and farmer perspective. *Bio-Based and Applied Economics*, *1*(1), 47–63.

Schütte, G. (2017). What kind of innovation policy does the bioeconomy need? *New Biotechnology*. PMID:28458016

Schwab, K. (2017). *The fourth industrial revolution*. Crown Business.

UNIDO. (2017). Accelerating clean energy through Industry 4.0: manufacturing the next revolution. A report of the United Nations Industrial Development Organization.

Van Lancker, J., Wauters, E., & Van Huylenbroeck, G. (2016). Managing innovation in the bioeconomy: An open innovation perspective. *Biomass and Bioenergy*, *90*, 60–69. doi:10.1016/j.biombioe.2016.03.017

ADDITIONAL READING

Aguilar, A., Bochereau, L., & Matthiessen, L. (2009). Biotechnology as the engine for the Knowledge-Based Bio-Economy. *Biotechnology & Genetic Engineering Reviews*, *26*(1), 371–388. doi:10.5661/bger-26-371 PMID:21415889

de Besi, M., & McCormick, K. (2015). Towards a bioeconomy in Europe: National, regional and industrial strategies. *Sustainability*, *7*(8), 10461–10478. doi:10.3390u70810461

Egea, F. J., Torrente, R. G., & Aguilar, A. (2018). An efficient agro-industrial complex in Almería (Spain): towards an integrated and sustainable bioeconomy model. *New Biotechnology, 40*(Part A), 103-112.

McCormick, K., & Kautto, N. (2013). The Bioeconomy in Europe: An Overview. *Sustainability*, *5*(6), 1–20. doi:10.3390u5062589

Philp, J. (2018). The bioeconomy, the challenge of the century for policy makers. *New biotechnology, 40*(Pt A), 11.

Scarlat, N., Dallemand, J. F., Monforti-Ferrario, F., & Nita, V. (2015). The role of biomass and bioenergy in a future bioeconomy: Policies and facts. *Environmental Development*, *15*, 3–34. doi:10.1016/j.envdev.2015.03.006

Wang, R., Cao, Q., Zhao, Q., & Li, Y. (2018). Bioindustry in China: An overview and perspective. *New biotechnology, 40*(Pt A), 46-51.

Zilberman, D., Kim, E., Kirschner, S., Kaplan, S., & Reeves, J. (2013). Technology and the future bioeconomy. *Agricultural Economics*, *44*(s1), 95–102. doi:10.1111/agec.12054

KEY TERMS AND DEFINITIONS

Bio-Based Products: Products wholly or partly derived from biomass, including plants, trees, and animals.

Blue Economy: A sustainable use of oceans for economic growth.

Circular Economy: Economics that refers to taking production as a basis for conversion and recycling instead of use and destruction.

Food Security: The condition in which people always have physical, social, and economic access to enough safe and nutritious food. This food meets dietary needs and food preferences for an active and healthy life.

Green Economy: A resilient economy that provides a better quality of life for all within the ecological limits of the planet.

Knowledge-Based Bioeconomy: The process of transforming life science knowledge into new, sustainable, eco-efficient, and competitive products.

Renewable Biomass: Energy sources derived from organic matter, include crop, waste, soybean, and garbage.

Chapter 3
Urban Farming in Sustainable City Development

Magdalena Grochulska-Salak
Warsaw University of Technology, Poland

ABSTRACT

Urban farming is defined as building development for the production of plants for the needs of the local community. The presented issues concern the shaping of urban farms for the preservation of the environmental balance of urban spaces and connections concerning the coexistence of architecture and greenery in the city. This chapter indicates the possibility of shaping synergistic spatial systems by integrating urban space and buildings with an innovative production function—a municipal farm—that complements the functional structure of the city in connection with the shaping of public spaces and the greenery system. The pro-environmental architecture connected with technologies enabling the production of plants in buildings enables the integration of urban space, complementing the functional and spatial structure of the city. The implementation of new technologies enables the production of plants in hydroponic and aeroponic farm buildings. The urban farm is an element in planning the city's sustainable development.

DOI: 10.4018/978-1-5225-7958-8.ch003

INTRODUCTION

Urban agriculture may be defined as building development created to host the production of plants for the needs of the local community. The presented issues concern the shaping of urban farms in order to preserve urban spaces environmental balance and to preserve connections concerning the coexistence of architecture and greenery in the city. As indicated, there exists the possibility to shape synergistic spatial systems by means of integrating urban space and building development with an innovative production function- an urban farm, which complements the functional structure of the city in connection to the shaping of public spaces, as well as the greenery system. The pro-environmental architecture connected with technologies enabling the production of plants in buildings makes integration of urban space possible, thereby complementing the functional and spatial structure of the city. The implementation of new technologies facilitates the introduction of plant production in hydroponic and aeroponic farm buildings.

The production technology, as well as the program and structure of the urban farm development provide the subject for research and analysis in order to shape building development integrated with greenery. As pointed out, it is necessary to increase the scope of production of tangible goods and intangible assets, which takes place in the urban farm complex. The connection with public spaces and the implementation of building development with a public utility function is justified by shaping the land development program and development for the needs of the local community. The urban farm is an element in planning the city's sustainable development. Research indicates that there exists the need to implement plant production and new technologies in the developing city and urban areas. The development of innovative technologies for plant production, energy acquisition and optimization of water consumption makes it possible to create urban farms within the city structure by means of integrating greenery with buildings. Eco-production arranged in multifunctional buildings is an architectural and urban planning issue from the scope of city development and revitalization of degraded areas. The urban farm is analysed in the context of the conversion of ways for supplying the local community with useable vegetation and food, as well as in the context of a special architectural task. The implementation of urban farms combines two trends towards shaping the city growth and city building development. The first trend may be referred to as a "smart city" that uses modern technology in the city structure. The second trend comprises pro-environmental solutions and "green architecture". The introduced solutions enable optimization and balancing of energy consumption, minimization of environmental interference and integration of greenery with both buildings and urban space. The construction of urban farms while planning the city growth should be connected with the systems of road transport, public spaces and urban greenery. While planning a balanced development of a city, the construction of urban farms in the social, production, architectural, economic and environmental context should also be taken into account.

The issue of city sustainable development is both multithreaded and interdisciplinary. The issue provides the subject for research, analysis and studies that constitute guidelines for designing and balancing areas, functional structures, building development and infrastructure in the city. The study presents research work results concerning the contemporary ways for shaping of the city, the development and transformations determinants, as well as urban farms in the city structure.

The study provides an elaboration on and a justification for formulated theses:

1. The concept of contemporary cities sustainable development is implemented owing to new spatial models, technologies and social functions.
2. The urban farm is implemented in a modern city for the sake of production function in order to fulfil the needs of the local community and for the sake of shaping the living environment parameters.
3. The urban farm is an element of the city's sustainable development model, it integrates spatial and architectural solutions, provides green infrastructure and eco-production in the city.

In order to present the research work and introduce evidence, the study was divided into three parts for the study structure:

1. The issue of city sustainable development - introduction to the issues of sustainable development and the contemporary model of city development.
2. Urban farm - production for the needs of the local community and city infrastructure
3. The urban farm as an element of the sustainable development model - conclusions summarising the research work evidence

The Issues of Sustainable Development of the City

At the end of the 20th century, a period of research and introduction of a new idea for the civilization functioning and development, namely the Sustainable Development Concept took its beginning. The definition of sustainable development dates back to 1987 and may first be traced back to the UN report entitled *"Our Common Future"*.

The publication resulted from the work of the independent World Commission on Environment and Development established in 1983. Indications of the new issues and threats to the modern world were included in the report. It has become the basis for the research and analysis development concerning optimization and progress directions in all economic sectors. The issue of sustainable development also concerns shaping the functional program, spatial structure and infrastructure of the city. The conditions resulting from the need to optimize energy acquisition and consumption management affect the architecture and solutions in construction industry. The idea of sustainable development is continuously improved and redefined with the passage of time and with the emergence of new studies and research results obtained in the fields of construction, engineering, sociology, economics and law.

Research on the continuation of sustainable development concept finds application in the society functioning model, in new technologies implemented and in studies for legislation purposes. The genesis and characteristics of sustainable development have been described in, among other publications, "sustainable development - global challenges" (ed.: Trzepacz.P, 2012). Development issues are the subject of international studies in which the principles of cooperation, co-action and guidelines for development strategy, spatial policy and economic mechanisms and systems in the country are established. The issue of sustainable urban landscape shaping and functional structure shaping in the context of improving the quality of life was indicated documents, such as *the Leipzig Charter on Sustainable European Cities* of April 27, 2007.

The issue of sustainable development in the European Union is described, among other documents, in the Europe 2020 Strategy - a strategy for smart and sustainable development conducive to social inclusion (Commission Communication COM 2010, Brussels 2010), as well as in a detailed study from 2015 entitled *Sustainable Development Goals* included in the 2030 *Development Agenda Transforming Our World* The development issues concerning the following the sectors of employment, innovation, education, social inclusion, climate change were indicated therein. The abovementioned areas of tasks specify the necessity for spatial, economic and social interdisciplinary activities. In the report entitled *Our Common Future*, it has been indicated that *in the process of transformation, which ensures meeting the needs of the contemporary generation, while not diminishing the development opportunities of future generations, thanks to such activities as integrated activities in the field of economic, social and environmental development.* (Report from the World Commission of G.H Brundtland for Environment and Development, *Our Common Future*).

In studies concerning the sustainable urban environment co-creation, research areas from various fields, as well as their interrelations are indicated. Ideas regarding the implementation of innovative technologies for urban equipment and the development of the "smart city" concept are emerging. The Smart City concept is related with the development and implementation of information systems connected to urban space management and analysis of environmental and social parameters. The idea of a smart city has been in progress for two decades now (Schaffers et al 2011). The smart city is a city in which the integration of human and technological capital takes place in order to improve the quality of life and to stimulate economic growth based on optimization in terms of natural resources management and participatory management - ICT - Information and Communication Technology, (Caragliu, A., et al 2009). Certain areas of a smart city development that may be referred to as smart include management, economics, mobility, environment, society, housing. The publications present the examples of 3 generations of Smart City (Cohen 2015). The first generation that may be distinguished, smart cities 1.0 were based on management for which information technologies were applied, but which lacked interaction with residents (Cohen 2015, Townsend 2013).

The second generation of smart cities, smart city 2.0 can be defined by integration of public administration activities and ITC implementation in order to solve local functional and environmental problems. The third generation, smart city 3.0 model is characterized as a city of social participation. In this model, residents and users are supposed to actively participate in the creation and in making use of information infrastructure and urban environment parameters. Publications and theoretical considerations present various models of urban development. Pro-social and pro-ecological solutions are described as desirable directions for the development of urbanized areas that would be consistent with the eco-development trend. Towards the end of the 20th century, the concept of Eco City was wide-spread, such as the one created by the Urban Ecology organization, (beginning Richard Register's activity, 1975). The basic issue was to analyse the city development by means of reconstructing and maintaining balance in the urban environmental system. By the end of the 20th century, the Eco City modelling criteria had been defined (Roseland 1997, Harvey 2011).

Postulates for the implementation of a self-sufficient economy and the usage of local resources were indicated. The resource management method is supposed to minimize the emission of carbon dioxide in production of energy from renewable sources. The Eco City concepts point to the need to save water and energy resources and optimize its usage. Included in Eco City models under development and in conjunction with the Smart City concept, such methods nowadays indicate the directions for development of a modern sustainable city constructed in such a way as to respect environmental, social, spatial and economic conditions.

Cities are analysed in reference to implementation the Smart City concept. In 2017, the IESE Cities in Motion Index 2017 Report was published. It shows that two Polish cities, Warsaw and Wroclaw, have been qualified among the first one hundred in the ranking of smartest cities in the world. The report highlighted New York, London and Paris in the lead. The ranking adopted guidelines in which a smart city is understood as a city that uses advanced technologies to improve the quality of life in all areas of its functioning. The IESE Cities in Motion Index 2017 Report takes into consideration all aspects that make up the quality of life in 180 cities around the world. The cities were analysed using a set of indicators related to human capital, social cohesion, economy, public management, social participation, the natural environment, mobility and transport, spatial planning, efficiency and advanced technologies. (businessinsider materials).

Research and investments are conducted so as to implement the concept of sustainable development, as well as smart and eco the cities. Strategies and studies on spatial policy for regional and local development are being developed. Polish scientists, drawing from global research and experience, indicate development directions and conduct their own research concerning technological advancement. Research is being carried out for the sake of implementation of technologies for monitoring and controlling urban environment and urban landscape parameters. Moreover, in Poland scientists conduct research for the city development and point to numerous spatial, functional, social and economic conflicts. Polish universities create and develop technology parks and "living lab" urban units, in which scientists have an opportunity to conduct research, implement new technologies and test them in everyday use. An example of such an investment is the town of the future implemented by the Warsaw University of Technology. The New Technologies Campus is implemented at 300,000 m2 of the area of the Warsaw University of Technology. The New Technologies Campus of the Warsaw University of Technology is planned as urban units, in which there will be residential, office and service buildings, all public spaces will be equipped with modern technologies and intelligent solutions, among which there will be plants capable of indoor air conditioning, smart lighting management and smart ventilation management, and sensors signalling the level of pathogenic bacteria in the room. The work has been under way since 2015, involving 100 companies that have already been implementing around 70 research and development projects, with the cooperation of 62 scientists from 9 faculties of the Warsaw University of Technology, together with 100 experts from various companies dealing with technology. 5 research areas related to everyday life in the city were selected: smart urban spaces, smart buildings, smart housing, industry version 4.0 (shortening and increasing efficiency for the production of the future), social changes related to the development of smart technologies. (businessinsider materials).

While defining development as the inevitable and necessary changes and transformations, the issue of sustainable development should be an optimal and effective set of actions applied to improve the urban environment in connection with the development of social and economic structures, with the use of civilization achievements. The analysis of the links between the causes and effects in relation to impact on the environment and the analysis of the needs di splayed by the contemporary society is aimed at shaping the multi-criteria model for the society and urban space development. The current definitions of sustainable development present a development model as a strategy and executive elaborations for the implementation of such solutions that would lead to activity optimization in various economic and social sectors. Interdisciplinary studies on sustainable development in Poland are published. The Polish collective publications "Sustainable city - ideas and realities" should also be mentioned as important (edited by: Jnuchta-Sostakak A., Banach M., 2016). Therein, design issues are described in spatial, architectural, social and construction aspects. One of the issues indicated for further implementation in spatial planning is the regenerative design of urban ecosystems in the urban environment.

A human being should become a part of the city's ecosystem and act in accordance with a deep respect and understanding of its processes (Wahl 2016, van der Ryn 2013). The concept of "city sustainable development" shows many meanings, interpretations and research areas. The author of this chapter has for years been dealing with the issues of sustainable development and design in the city. In the publication from 2011, it was stated: *It can be assumed that the basic model assumptions remain constant, however the way they are shaped and developed varies in time depending on technical and cultural progress. An approximate description of an effective city model shaped in a regular, correct, balanced way should include spatial answers that implement development tasks* (...) (M. Grochulsa-Salak, 2011). Among the tasks to be implemented in the development of city functional and spatial structure, the issues identified include: security, public spaces, communication, housing and private spaces, services and multi-functional units, infrastructure, industry, public administration, cultural and art facilities, ethics and morality objects, education, health care. The publication also pointed to the importance of interactions between urban tissue development and technological achievements.

In the city, impulses and stimulators of social, technical, economic and ecological development arise, at the same time urban tissue and architecture become recipients of these achievements. The development of thought and technology affects the transformation of space and building developments. (M. Grochulsa-Salak, 2011).

Cities treated as a "living organism" are subjected to changes and transformations as a result of advancements in thought and technology. Therefore, ecological, economic, spatial and political incentives should define the development of a city treated as a complex organism, equipped with information systems and management methods.

The idea of optimization and analysis of impact effects determine the implementation of new solutions consistent with the needs of the local community. The principles of the city-garden, smart city and Eco-city can be considered the basis for constructing city sustainable development concept. Development according to the principles of optimization. Constructing in that way would be balanced and interdisciplinary in a qualitative, aesthetic, social and economic context. The implementation of city sustainable development concept is related to the change in lifestyle and the way of thinking of contemporary and future generations. Analysis of the existing state, forecasting and optimization and implementation of solutions in the multi-criteria model. The analysis of theoretical works provides grounds on which to assert that the idea of "sustainable development" concerns the shaping of urban space and the urban environment with the use of new information techniques and methods for shaping the urban ecosystem. The implementation of the sustainable development assumptions in the city introduces green infrastructure into the city and the production of tangible goods and intangible assets to fulfil the needs of the local community and related urban centers. In UN studies, sustainable development is presented in three aspects: ecological, social and economic. The implementation of a urban farm concerns all the three indicated three issues. Namely, Eco-production in the city, which is a case integrating the issues of environment shaping and implementation of pro-environmental technologies; social changes and services for the local community; economic conditions for the implementation of multifunctional investments with production in the city.

The redefinition of production is related to the recommendation to minimize contemporary material, non-material and non-renewable resources consumption. Therefore, the idea of sustainable development concerns the decision-making process and changes in the functioning of the local community in connection with social, economic and spatial consequences.

Nowadays, a city farm can provide an element that is bestowed with the ability to integrate the indicated issues. It can be treated as an element of the city-garden concept and smart city concept development in the process of city sustainable development. Therefore, an urban farm is the topic selected for further analysis in the context of shaping a sustainable city.

An Urban Farm: Production for the Needs of the Local Community and City Infrastructure

The design of integrated green infrastructure in the city is one of the novel, modern city design issues. Greenery was gradually (since the nineteenth century) introduced to urban spaces, at first as a continuous compositional element (alley) or as a nest compositional element (garden/park). The idea to shape such city in which it was possible to introduce the model of the city garden described and created by Ebenezer Howard (the city model was shaped in economic and spatial terms) provided another stage of incorporating greenery into city structure. The garden city concept was developed and interpreted by successors. Cities and districts that implemented the urban idea of the "garden city" were constructed. Greenery was introduced due to its beneficial hygienic and aesthetic qualities in the city. However, in the model of a city-garden, the introduction of greenery was not the essence of matter. In its original assumptions, models of the city-garden idea concerned the coexistence and mutual penetration of urban functions and rural services and production. The idea of a "social city" and a functional "city-village" was ahead of technological possibilities, as well as of social and economic transformations.

Only recently has it become possible to implement city-garden, with production and cultivation in the city structure, in an efficient, economical and socially accepted manner. Contemporary trends in the design of the city structure introduce greenery treated as biomass. Greenery is not only a compositional and aesthetic element. Greenery in the city becomes a functional element of the city equipment. Biomass and green infrastructure have a beneficial effect on the environmental parameters and microclimate in the city. Greenery exerts also a positive impact on people's functioning and social relations. The implementation of greenery has an influence on integrating the local community. The analysis of modern green infrastructure implementation in the city indicates the function of regulating the parameters of the urban environment in the city, temperature, humidity, ventilation, biodiversity and ecosystem efficiency. The biomass production for energy purposes and for food production are also added to the tasks listed nowadays. Food production and biomass productions (to be processed) in the city occurs in Urban Farms.

The issues concerning greenery in the city in the context of sustainable development determines research on the feasibility and form of vegetation in the city. Design and research works are carried out on biomass usage and functionality in the city.

The author of the study conducts research on urban farms and on implementation of buildings integrated with urban greenery. The results of scientific work, literary analysis and ongoing investments have been described in previous publications, among which the publication such as. "Shaping buildings and greenery systems in urban areas" (Grochulska-Salak M., Zielonko-Jung K., Zinowiec-Cieplik K., 2018) may be listed. The study points out the coexistence and integration of buildings and green infrastructure. In the conclusions yielded by the interdisciplinary research team, the proposals for hybrid urban units have been described. In the part concerning urban farms, a typology of urban farms to fulfil the needs of food production in the city is described. As a result of the research work, the typology of plant cultivation in the city was described. The first type of typology is the cultivation of plants planted in greenery. Most often such farms are implemented on a smaller scale in the open space or on the roofs and on the walls of buildings. These crops are not protected against environmental contamination and weather conditions. Examples of such farms are: social and private gardens in the city on the home soil or in pots with soil. In the case of vegetation cultivated on roofs or building walls, irrigation management systems are necessary. In figure 1, an example of a farm on the roof of the RK / Riverpark Farm office building is presented.

Another category of crops is the semi-automated production of indoor plants. Vegetation in pots in the ground or properly selected substrate. It is necessary to apply technical solutions that control and implement adequate hydration, lighting and environmental

Another category of typology is the semi-automated production of indoor plants. Vegetation in pots in the ground or in properly selected substrate. It is necessary to apply technical solutions that control and implement adequate hydration, lighting and environmental parameters inside the building. Through the use of smart solutions, it is possible to increase the effectiveness and efficiency of production protected from pollution of the external environment and pests.

The Kono design implementations may serve as the example. They include such realizations as Pasona Group office building, constructed in Japan, Tokyo 2010 and Pasona O2 constructed in Tokyo in 2005 (the portfolio of konodesigns.com), presented in figure 2.

Another described type of plants is the industrialized production of plants in buildings in the city. Fully automatic, highly specialized cultivation, in which the vegetation and its growth parameters are controlled by means of adjusting the

Figure 1. Farm on the roof of the RK / Riverpark Farm office building

Figure 2. Pasona Group office building in Japan, Tokyo

environmental, lighting and nutrient parameters contained in the water in which the plants are rooted. Depending on the adopted technology, hydroponics (in liquid water), aquaponics (in water using fish) and aeroponic (a mixture of spraying minerals in the root zone of plants) techniques are distinguished. The hydroponic crop is illustrated in

Technological possibilities and ideological assumptions for the introduction of plant cultivation inside city buildings provide an impulse for redefining the city. Considerations regarding creation of city sustainable development are closely connected with the new formation of the functional and spatial structure, as well as greenery and buildings integrated with biologically active areas.

The issue of urban agriculture was also analysed in the context of suburbia development. Social, economic and spatial issues concerning implementation of urban farms in urbanized areas are being indicated nowadays. In the summary to the article, which indicates technologies that enable the integration of greenery with buildings, multifunctional buildings with plant production can be a spatial dominant and a landmark. The author states: *Introduction of urban agriculture through the implementation of multifunctional urban farms connected to the public space at*

Figure 3. Hydroponic and aeroponic farm

develop technologies for a private investor, who announced in 2015 the construction of an algae farm in the Lubuski Industrial and Technological Park. A farm will be created for the production of electricity from biogas resulting from the fermentation of algae (Szulc G. 2015).

28 covered reservoirs with an area 3.800 m2 for the production of algae biomass, for use as a biogas plant. As a result of conducting anaerobic digestion and biogas combustion in gas engines, electricity and heat will be generated. 1 ha of algae gives approx. 300 tons of dry matter. Biogas produced during the fermentation of algae will be used to produce electricity. The digestate, rich in water and nutrients, will be used for the production of algae.

At the Warsaw University of Technology, another research is in progress, namely research on the breeding and use of algae for the conversion of carbon dioxide (after separating it from the fumes) and for the production of biomass and fatty acids used for the manufacture of biodiesel. A prototype installation for microalgae cultivation, a special strain, isolated in Germany (Halle, Saxony) was developed. Breeding takes place under pressure up to 10 bar. Temperature, pressure, exposure and pH parameters are monitored. The apparatuses are mobile, have its own source of light and gas. It is also possible to take samples for periodic testing of the biomass concentration of microorganisms. (materials from Warsaw University of Technology). Research on four categories in the typology of cultivation within the city, namely algae breeding, indicates the possibility to apply the farm for the purposes of various sectors of industry, especially the energy sector, and for improving microclimatic conditions, including factors such as oxygenation and air purification. The algae farms and their application may provide a very important element of the economy and the functional and spatial structure of cities in the future. These types of farms will therefore constitute a vital element of the city sustainable development and should be taken into account while conducting planning studies and research on green infrastructure and biomass production in the city. So far, in cities, green infrastructure has been implemented in order to improve the quality of life and the parameters of the urban environment. Vegetation as biomass integrated with buildings, is implemented in the form of roof gardens, vertical gardens and elements of the road system equipment. In 2017 in Mexico, the Via Verde project was implemented, which introduces greenery to road infrastructure. The investment concerns one thousand vertical plantings, which are intended to wrap the pillars of the city road beltway. As a result of this application of greenery, 60,000 m2 of green planting will appear, which will absorb heat and rainwater and purify the air. It is assumed that 27 thousand tons of toxic gases will be filtered annually, the biomass is to capture up to5 tons of suspended dust and 10 tons of heavy metals. The "Lowline" project, presented in 2012, comprises a park growing in the underground of Manhattan, in the intensive building development of the Lower East Side. The Lowline Park is to be located in the place of underground railway station, unused for the last 60 years, that used tooperate between Brooklyn - Delancey Street Station. In order to deliver daylight to vegetation, the designers invented a cup-shaped construction that collects light from above the ground, sends it down with the use of fiber-optic cables, so as to give the light back to the underground trees above the crowns of park. The venture may be seen as a part of the growing trend for underground construction.

The issue of urban farms does not only concern the cultivation and production of vegetation. The problem includes animal breeding in the city to cater for the needs of the local community. This issue in Poland is dealt with by researchers, such as Justyna Kleszcz (Kleszcz J. 2016). The issues concerning development of the Agropolis model in Poland are, furthermore, described by Bogusław Wowrzeczka (Wowrzeczka, 2014). In modern cities, one can observe the phenomena of urban farms creation. Such farms provide biologically active areas in the urban space and are designated for the purpose of cultivating useable vegetation as food (Herman, Kassner 2014).

The analysis of literature, research and investment works points to the fact that the issue of shaping urban farms is a scientifically recognized subject. Examples of urban farms are published as innovative and visionary solutions in the urban space. The implementation of urban farms in the city structure is an issue that needs to be recognized in the context of future challenges posed by spatial economy, that is the challenges of the projected sustainable development in relation to the functional and spatial structure of the city. Research works concerning the potential and implementation of urban farms indicates the complexity of the problem. The urban farm need not only be seen a green area within the city, a biologically active surface, or biomass for processing. Designing an urban farm is a matter of economic strategy, urban planning, architecture, sociology and economy.

The rationale for implementing green areas in the city and the ways to do it are diverse. *These factors include: changes in political and economic conditions on a global and a national scale, social needs resulting from economic opportunities, organization of investment processes and the character of such investments, organization and level of design and performance, and the general level of culture. The impact of global tendencies in the development of architectural thought provides a very important factor as well, so does the influence of social opinion and mass media that has not been appreciated enough as yet. (Wejchert K., 1984)* The development of the idea to model social and economic structures affects the creation of space and buildings in the city. Urban design as an area dealing with shaping human life environment in the process of urbanization, thus forming social, economic, spatial and construction structures, fulfils analytical and prognostic tasks. *In urban planning, we mainly deal with spatial development, that is advancement in building development, which, as a result of this process, obtains certain advantages more beneficial from the user point of view. (Chmielewski J. M., 2010)*

To conclude, the analysis of social, economic, technological and spatial transformations points to the necessity to consider construction of ecological cities in such a way as to take into account the possibility to cultivate useable plants there. This neds to be done with the application of new technologies in shaping buildings and areas of cultivation.

The system of urban greenery and biologically active areas should be supplemented with the function plant cultivation in order to fulfil the needs of local communities, in accordance with contemporary trends for urban areas transformation. It is advisable to continue the research work concerning the analysis of technological and architectural solutions for the implemented building developments that would be beneficial to the production of useable plants, algae and biomass. The implementation of urban farms is determined by political and economic decisions regarding food security.

Due to the introduction of urban farms, factors such as functional program, landscape and ecosystem of the city change. In this study, examples of shaping farms and green urban infrastructure have been briefly presented. Spatial and functional solutions for urban farms display a variety of possibilities. A typology of urban farms has been described above. These farms are implemented as:

- Green areas and social cultivation areas created in open areas in the city
- Gardens and biological areas on the roofs of existing buildings
- Vertical farms on the walls of buildings
- Development of underground structures and urban infrastructure
- **"Revitalization":** Plant cultivation in existing post-industrial buildings
- **"Eco-Production":** Creation of new buildings for the needs of industrial production with increased crop efficiency
- Building equipment and supplementing the functional program of multifunctional building developments, together with accompanying infrastructure

The introduction of urban farms is an element of sustainable development characterizing the changes that are observable in the spatial and economic structure of the city. The concept of urban farms integrates three ideas for the development of contemporary cities: Smart City, Eco City and City-Garden.

Urban Farm as an Element of the Sustainable Development Model

The issue of sustainable development and harmonious shaping of urban development is a city-planning and architectural issue. Collaboration between scientists from various fields and research by interdisciplinary teams that develop the initiated ideas and implement new technologies in the city structure is required..

It is currently possible to observe a far-reaching transformation of the environment by the human population. These changes are of global character. In some situations, they have gone so far that places that humans have inhabited for a long time have become dangerous to live in. This danger results from soil poisoning, water and air pollution.

This creates conditions in which the population becomes susceptible to an increasing number of diseases not only of physiological but also biotic (viruses, bacteria) background.

... Most of the resources that constitute the material premises for the current civilization development are non-renewable, which further influences the nature of the existing transformations.

(...) Awareness of a new way of thinking seems to be getting more and more common. Śpiewakowski E. R., Korczyński M., 1999)

Modern trends for technological development allow for re-introduction of production into cities. Eco-production in urban farms uses vegetation and biomass to benefit the local community and to improve environmental parameters, which provides a forecasted element of sustainable development.

Urban planners discuss the necessity of forecasting and noticing connections between the technological development, civilization and spatial development of city structures. *Familiarity with the laws governing development in the modern world becomes indispensable, because, on the one hand, it allows the application of these laws as tools for development control, while, on the other hand, it makes it possible to register developmental threats and causes of the emerging developmental tensions.*

(Chmielewski J. M., 2010). It is indicated that the case does not simply concern spatial planning, urban planning or architecture, but is related to the ability to build the world of the future. (Gzell S., 2015)

Human activity understood as civilization progress is related to the character of the transformation occurring in the living environment, to human impact on the environment in connection with the development of skills and technologies. *From the very beginning, there was a visible tendency to focus societies in cities that have become new ecosystems. Cities currently host most people in developed countries. (...) These ecosystems had their host zone, which, in historical times, was often defined in the acts granting city rights, but which zone is currently designated by an array of socio-economic links.* (Śpiewakowski E. R.,. Korczyński M, 1999) The process of urbanization is connected with political and economic, technical, cultural and ideological development. The urban farm is a advancement in the city garden concept and is inscribed in the trend towards the concept of city eco development. With the implementation of innovative technologies, green infrastructure in the city serves as the implementation of the "smart city" assumptions.

In the study, a typology of urban farms was described, while current tendencies for implementation of vegetation (biologically active areas and biomass, described above) that would be integrated with buildings in the city were indicated. The above-mentioned studies show the importance of current and future implementation of urban farms and plant production in the city.

An urban farm is a multi-faceted issue. The cultivation of vegetation in the city is of aesthetic and functional importance. Such vegetation serves as food, to shape environmental parameters, as a source oxygen production, as biomass for generating electricity, heat and biofuels. Implementation of farms as green infrastructure in the city is of economic importance (development of new employment and production sectors, striving for economic self-sufficiency), economic (balancing expenditures to material and energy gains, optimization of production, transport, infrastructure), social (services for the local community, production food, new places of employment, spaces of social integration, recreation and education) and environmental (improvement of the urban environment parameters, supplementation of the city natural system and improvement to ecosystems), climate (air purification, absorption of heat and water, obtaining fuels and energy from renewable sources using biomass). It may therefore be said that an urban farm applies to all aspects of Sustainable Development. Further development and introduction of cultivation in urban tissue is an issue of urban and architectural character. This fact justifies the claim that a city farm should be treated as an element of city sustainable development. This assumptions determines the implementation of urban farms in the functional and spatial structure of contemporary cities. It also points to the necessity to conduct further research concerning the implementation of urban farms in urbanized areas. Eco-production and green infrastructure should be created if a harmonious and effective city development is to be achieved. Transformations in the urban tissue, as well as in functional and spatial structures of cities should be introduced with great care so as not to disrupt the identity of the place. The proposed model hybrid urban units introduce city farms and green infrastructure integrated with public utility buildings. A city development that is both harmonious and sustainable should take into account the foregoing achievements of civilization with due respect to history and continuity, also in the field of architecture and urban planning.

REFERENCES

Chmielewski, J. M. (2010). *Teoria urbanistyki w projektowaniu i planowaniu miast*. Warszawa: Oficyna Wydawnicza PW.

Cohen, B. (2013). *Smart City Wheel*. Retrieved from https://www.smart-circle.org/smartcity/blog/boyd-cohen-the-smart-city-wheel/

Cohen, B. (2015). *The 3 Generations Of Smart Cities. Inside the development of the technology driven city*. Retrieved from https://www.fastcompany.com/3047795/the-3-generations-of-smart-cities

Grochulska-Salak, M. (2011). Projektowanie urbanistyczno-architektoniczne w mieście uwzględniające zasady zrównoważonego rozwoju w Energia i Budynek 02(45)201, Warszawa 2011.

Grochulska-Salak M., (2017). Rolnictwo miejskie jako element krystalizujący przestrzeń suburbiów. *Kwartalnik Naukowy Uczelni Vistula, 4*.

Grochulska-Salak, M., Zielonko-Jung, K., & Zinowiec-Cieplik, K. (2018). Prace Naukowe Uniwersytetu Ekonomicznego we Wrocławiu nr XXX: Gospodarka przestrzenna – stan obecny i wyzwania przyszłości. Academic Press.

Gzell, S. (2015). *Wykłady o współczesnej urbanistyce*. Warszawa: Oficyna Wydawnicza PW.

Harvey, F. (2010). Green vision: the search for the ideal eco-city. *Financial Times*. Retrieved from https://www.ft.com/content/c13677ce-b062-11df-8c04-00144feabdc0

Kleszcz, J. (2016). Farma w mieście – wizja rolnictwa XXI wieku. *ARCHITECTURAE et ARTIBUS, 3*. Retrieved from http://www.wa.pb.edu.pl/uploads/downloads/Architektura-3-2016---artykul-6--do-internetu-.pdf

Roseland, M. (1997). Dimensions of the Eco-city. *Cities (London, England)*, *14*(4), 197–202. doi:10.1016/S0264-2751(97)00003-6

Schaffers, H. (2011). Smart Cities and the Future Internet: Towards Cooperation Frameworks for Open Innovation. In *The The Future Internet*. Future Internet Assembly, Springer. doi:10.1007/978-3-642-20898-0_31

Śpiewakowski, E. R., & Korczyński, M. (1999). *Rozwój cywilizacji ludzkiej w kontekście sukcesji ekologicznej, „Hipoteza ekologii uniwersalistycznej",Centrum Uniwersalizmu przy Uniwersytecie Warszawskim Polska Federacja Życia*. Warszawa: Drukarnia Jerzego Kosińskiego.

Szulczewska, B., & Bruszewska, K. (2013). *Urban Agriculture in the city of Warsaw*. Poland: Academic Press.

Szwed, D., & Maciejewska, B. (2011). *Zrównoważony rozwój metropolii*. Gdańsk: Fundacja Przestrzenie Dialogu.

Trzepacz, P. (Ed.). (2012). Zrównoważony rozwój – wyzwania globalne. Podręcznik dla uczestników studiów doktoranckich, Instytut Geografii i Gospodarki Przestrzennej UJ.

Wejchert, K. (1984). *Elementy kompozycji urbanistycznej*. Academic Press.

Zinowiec-Cieplik, K. (2017). Potencjał integracji form roślinnych z architekturą - środowisko i technika. *Kwartalnik Naukowy Uczelni Vistula*, *4*(54), 2017.

KEY TERMS AND DEFINITIONS

Eco City: Concepts of city that need to save water and energy resources and optimize its usage and management method is supposed to minimize the emission of carbon dioxide in production of energy from renewable sources.

Green City: Comprises pro-environmental solutions, optimization and balancing of energy consumption, minimization of environmental interference and integration of greenery with buildings and urban space.

Smart City: Is a city in which the integration of human and technological capital takes place in order to improve the quality of life and to stimulate economic growth based on optimization in terms of natural resources management and participatory management.

Sustainable Development: Inevitable and necessary changes and transformations for optimal and effective set of actions applied to improve the urban environment in connection with the development of social and economic structures, with the use of civilization achievements.

Urban Agriculture: Building development created to host the production of plants for the needs of the local community.

Chapter 4
Parametric Evaluation of Beam Deflection on Piezoelectric Material Using Implicit and Explicit Method Simulations:
A Study in Energy Engineering

Rakesh Nath
https://orcid.org/0000-0003-2757-3442
Girijananda Chowdhury Institute of Management and Technology, India

ABSTRACT

Energy harnessing methods from rainwater using various smart materials have come into research, thus widening the scope of using these materials for the implementation in modern use. The piezoelectric materials present a brief idea of voltage generation whenever the material is deflected. The purpose of the study is to visualize an interconnection in parametric analysis of piezoelectric effect based energy harvester using two different commercially available piezoelectric materials, PZT-4A and PZT-5H, through series of implicit and explicit method simulations of FEM on COMSOL and ANSYS. The dynamic loads of different rainwater droplets sizes are investigated analytically. To calculate the variation of different methods in terms of deflection and voltage output, the implicit and multi-body explicit dynamic simulations are implemented separately.

DOI: 10.4018/978-1-5225-7958-8.ch004

INTRODUCTION

In the present world, the requirement of energy in our day to day life is rapidly increasing and so the necessity of production of energy by various alternative sources. Among the various alternative technologies rain water harvesting and its utilization are an environmentally sound and a very economical solution. Rainwater harvesting, in its broadest sense, is a technology used for collecting and storing rainwater for human use from rooftops, land surfaces or rock catchments using simple techniques such as jars and pots as well as engineered techniques. Rainwater harvesting has been practiced for more than 4,000 years, owing to the temporal and spatial variability of rainfall. It is an important water source in many areas with significant rainfall but lacking any kind of conventional, centralized supply system. It is also a good option in areas where good quality fresh surface water or groundwater is lacking. The application of appropriate rainwater harvesting technology is important for the utilization of rainwater as a water resource (Dos Anjos, 1998).

Piezoelectric materials are smart materials which are widely used to convert mechanical energy into electrical energy. Contacts are attached to the material; a quantity of electric charge proportional to the applied force can be measured. Piezoelectric effect occurs only in non-conductive materials (Ramsden & Dix, 2003).

Figure 1 shows the piezoelectric effect when compressed or stretched.

There is a tremendous amount of vibrational and kinetic energy which can be utilized by the falling of rainwater droplet on piezoelectric plate, which might be in the form a Bimorph viz. two active layers of one piezoelectric material and another that of structural steel or any other metal. It has been explained how the kinetic energy of raindrops can be converted into electricity by using piezoelectric materials (Biswas, 2009). This chapter explains the behavior of raindrops at impact and of piezoelectricity.

Background

The definition of smart materials has been referred to that class of materials that respond to multiple stimulus producing a useful effect that may include a signal that the materials are acting upon it. The important element of a smart material is the action of determining if a material is smart pertains to its asymmetrical nature. This is primarily critical for the piezoelectric materials. Materials having the label of being smart includes piezoelectric materials, electrostrictive materials, electrorheological materials, magnetorheological materials, thermoresponsive materials, smart catalysts, and shape memory alloys.

Figure 1.. Development of an internal electrical field by a piezoelectric material when compressed (A) or stretched (B) (Ramsden & Dix, 2003)

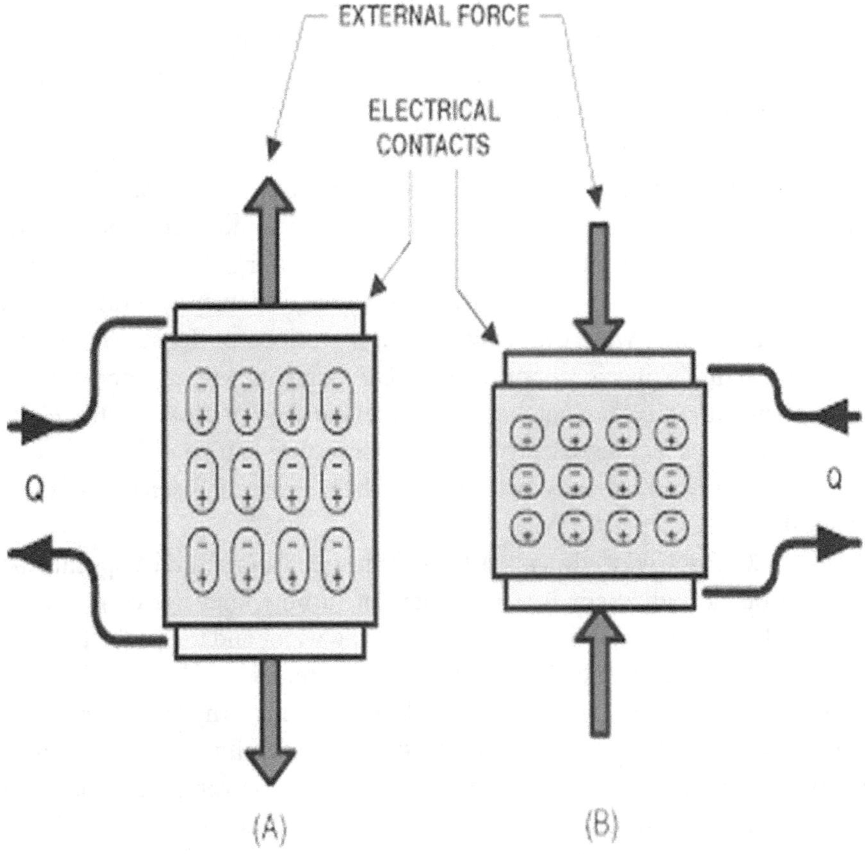

The piezoelectric effect was discovered when a mechanical stress was applied on crystals such as tourmaline, topaz, quartz, Rochelle salt and cane sugar, electrical charges appeared, and voltage was proportional to the stress applied. First of its kind were the piezoelectric ultrasonic transducers and soon swinging quartz for standard of frequency (quartz clocks). A refined example of those is the car's air bag sensor. The material detects the intensity of the shock and sends an electrical signal which triggers the air bag.

The author in this chapter has elaborately discussed the potential usage of Piezoelectric materials (Smart materials) for underlining the economical solution to a healthy living. The work was preceded by an extensive research on topics, primarily dealt with the deflections of the strained material. Popovici et al. (2008) studied the FEM analysis of a cantilever beam deformation producing electrical voltages through a direct piezoelectric effect of PZT-5H. Ahmed Telba (2012) developed a piezoelectric harvesting technology replenishing the minimization of the requirement of an external power source as well as the maintenance requirement for periodic battery replacement.

Pozzi (2011) developed Frequency up-conversion strategy of increasing the efficiency because the piezoelectric devices are permitted to vibrate at resonance even if the input excitation occurs at a much lower frequency. An experimental rig has been considered to observe the response of the bimorph in the harvester.

Makki (2011) concluded three different methods of power generation utilizing PZT and PVDF which were evaluated based on various performance criteria. A novel method using a plastic ribbon with PVDF elements was also presented as an alternative with attractive advantages over direct bonding of piezoelectric elements to the tire. In any case, piezoelectric energy harvesting can be effectively used as a substitute for limited capacity batteries in wheel embedded sensors to elongate their useful life.

Tinaikar (2013) introduced a system by making use of a structural disposal pipeline system, use of individual small scale generator turbine, and use of piezoelectric generator to harness the kinetic water of the falling water and the project deals with the required piping design needed for maximum output.

Martin (2013) examined the feasibility of designing a micro hydel power generation utilizing the harvested rain water at the top story of the building and another as the underground storage tank for collecting the water after power generation for other uses. The design of storage tanks pipe network and flow control valve will be done for optimizing utilization of the harvested rain water.

Biswas et al. (2009) explained that the kinetic energy of raindrops can be converted into electricity by using piezoelectric materials. The author explained the behavior of raindrops at impact and of piezoelectricity. The physics of monsoon and rain pattern are also described in detail. It has been found that Polyvinyldine Fluoride called PVDF is the most suitable type of material for power harvesting application.

A review of the published literature lends an insight into the importance of the piezoelectric materials and their design implementation. The choice of piezoelectric materials has been found to be an important part of rainwater harvesting demonstration, providing in depth solutions to the output obtained. The implicit and explicit dynamic study of the following will give an accurate and deep understanding at impact and concentrated zones.

Based on thorough literature review it is seen that most of the studies concentrate on selection of piezoelectric materials and the performance enhancement studies of the deflection of the strained material. The literature has also shown that many of the simulations, done successfully on ANSYS® and MATLAB, define better stability in terms of holding the tolerances to the extremities. The mesh structure is clearly refined, which tends to bring a comparative study of the literature to the authors's desk.

MODELING AND IMPLEMENTATION

In this section, the author described about the selection of various piezoelectric materials and their implementation. Piezoelectric plates have been considered in the form of rectangular cross section, which when impacted against the energy of the raindrop gets strained and generates electric charge proportional to applied force by the droplets. Thus mechanical energy is converted into electrical energy.

STRUCTURAL ANALYSES

Structural analysis is needed to determine the parametric evaluation of the beam undertaken. The general-purpose finite element analysis (FEA) software package COMSOL and ANSYS® Workbench were used for the design validation through implicit and explicit simulations of the piezoelectric cantilever beam. Popovici et al. (2008) have done a similar analysis of a cantilever beam using PZT-5H in COMSOL.

GEOMETRY OF THE MODEL

To simulate the structure, the author has chosen a multiphysics problem with PZT-5H and PZT-4A for the implicit and explicit dynamics simulation. As discussed earlier, Piezoelectric material study has been carried out in rectangular cross sectional plates which on deformations produces electric charge. The geometry used is presented in figure below:

Figure 2 shows the cross sectional of a geometrical configuration under study.

Table 1 shows the geometric details of the rectangular plate. The dimensioning has been kept true to the original as used by Popovici et al. (2008)

Popovici et al. (2008) incorporated two separate domains for their structural analysis. Domain R1 has been considered as an isotropic structural steel beam

Figure 2. The cross sectional view of a cantilever beam

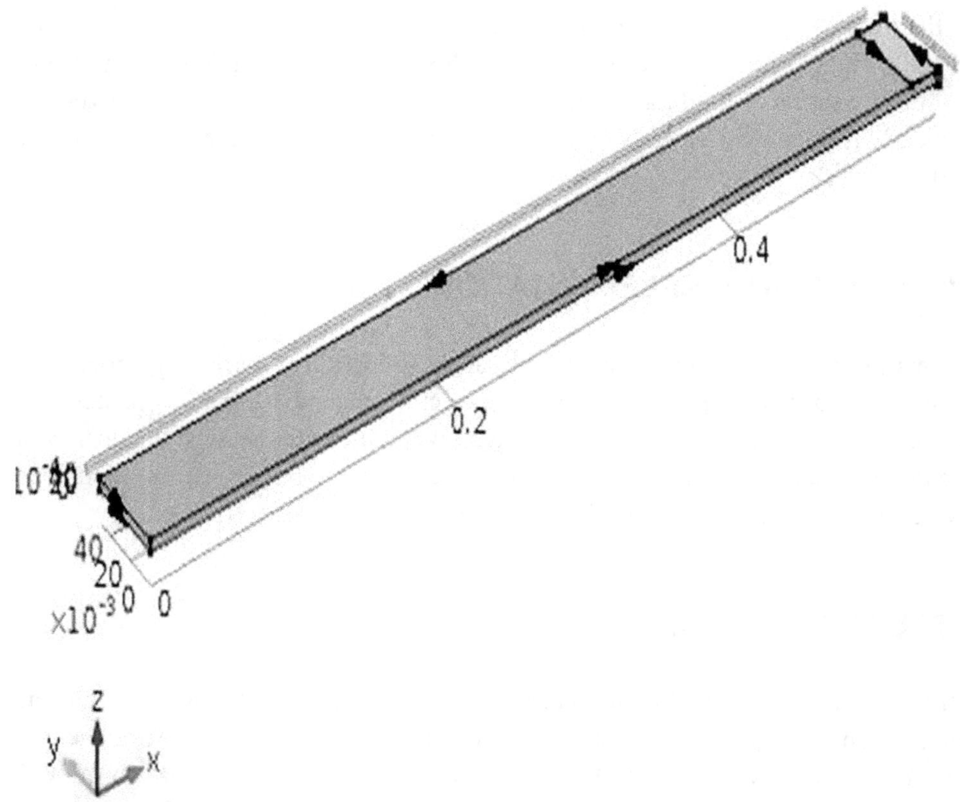

Table 1. Geometry details

Domain	Material(m)	Width(m)	Depth(m)	Height(m)
R1	Structural Steel	.55	.05	.005
R2	Piezo	.02	.05	.0005

with a length of 550 mm, width of 50 mm and thickness of 5 mm. This material is defined in Library1 (COMSOL,2005), while domain R2 is the PZT 5H cell which has a length of 20 mm, width of 50 mm and thickness of 0.5 mm.

Table 2 shows the material constants for the structural steel domain.

Popovici et al. (2008) considered the following parameters for PZT – 5H properties as those listed in (COMSOL, 2005).

Figure 3. The geometry of domain R1
(adopted from Popovici et al.,2008)

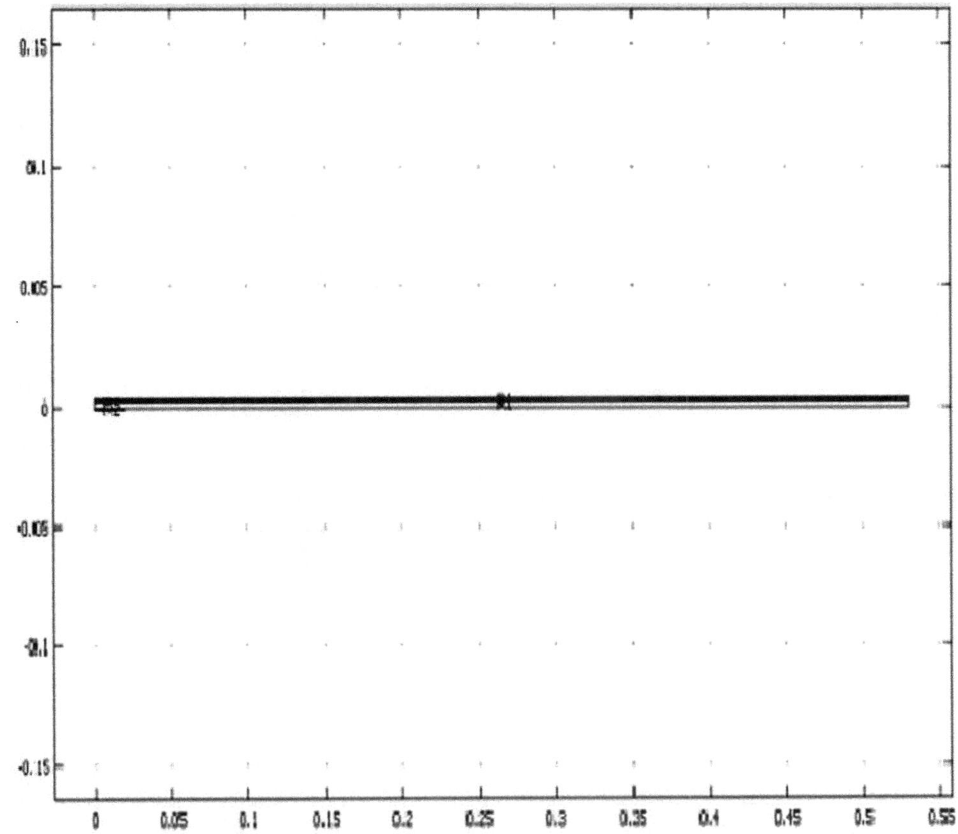

The values of the elasticity matrix are as such

c_E (c11=c22=126, c12=80.5, c13=c23=126, c33=117, c44=23.3, c55=c66=23, all in GPa)

Popovici et al. (June 2008, pp 479 – 481) also discussed about the values of the piezoelectric stress material, given by

e (e51= e42=17, e13= e23=17, e33= 23.3, [C/m^2]),

ε_S (ε11 = ε 22 =1704, ε 33 =1433, relative values).

Figure 4. PZT cell section of domain R2
(adopted from Popovici et al.,2008)

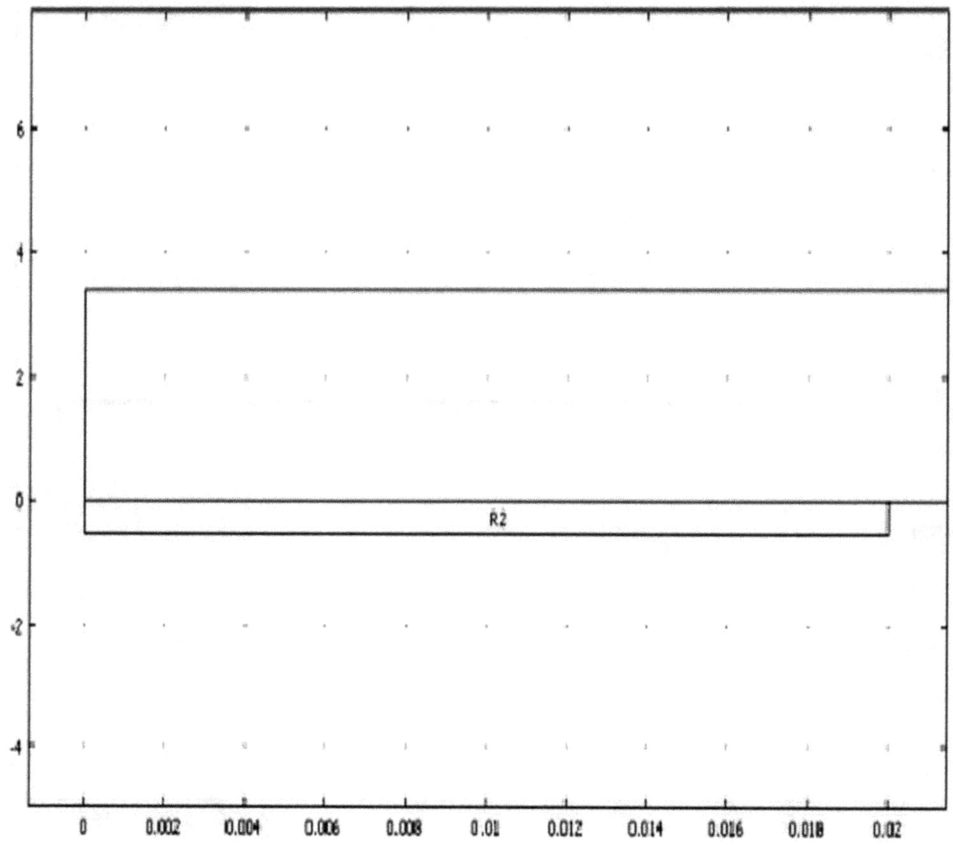

Table 2. Structural steel material constants

Material Constants	Value
Young's Modulus, E	2 x 10^5 MPa
Poisson's ratio, ν	0.33
Density, ρ	7850 kg/m^3

The boundary conditions for the mechanical part of the problem was formulated with a constraint of zero displacement on the left side of the beam with consideration of PZT cell. The load was applied on the right end of the beam only on the y direction.

Figure 5 shows the boundary conditions of the structure.

Figure 5. The mechanical boundary conditions
(adopted from Popovici et al.,2008)

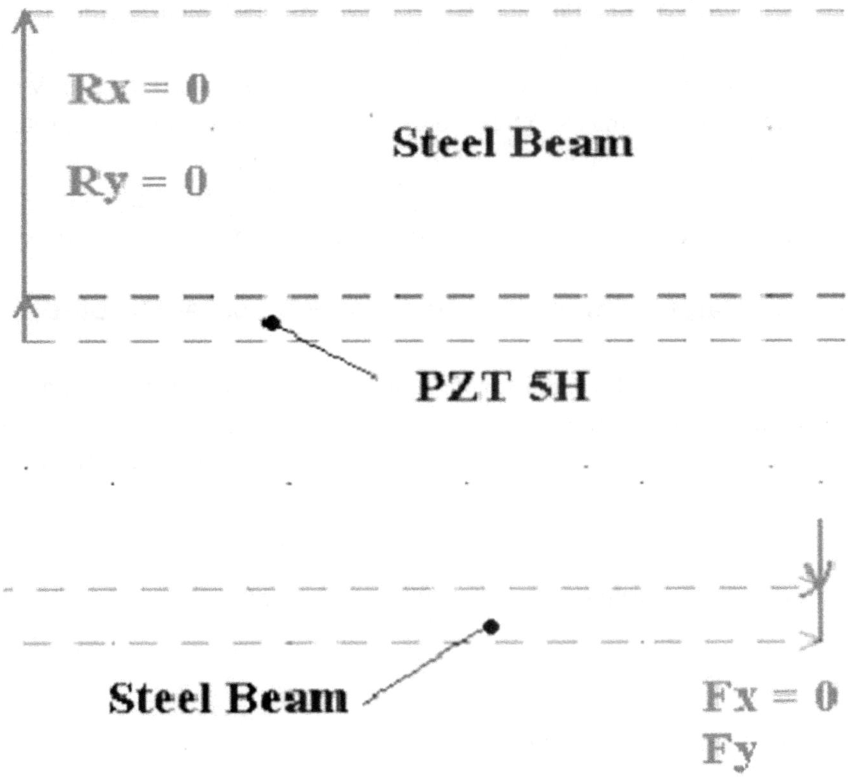

Popovici et al. (2008) in further claimed that the horizontal bottom surface of the PZT cell was set to ground and a zero charge/symmetry condition was set on the top surface.

SIMULATION STEPS

The author in this chapter dealt with simulation based on COMSOL, ANSYS® Static Structural and ANSYS® Explicit Dynamics to compare the results procured from Implicit and Explicit FEM simulation.

IMPLICIT FEM SIMULATION

Popovici et al. (2008) had established a 2-D FEM (Finite Element Method) analysis of a cantilever beam deformation producing electrical voltages through a direct piezoelectric effect of PZT in COMSOL. This chapter is referred to carry out an implicit 3-D simulation in ANSYS® and COMSOL with same material properties; dimension and line load, hence, validating the simulation.

Figure 6 shows the model and the mesh generation of the plates layered with PZT The mesh method was automatic and mesh generation type are free quad.

Analytical Calculation of Kinetic Load Rainwater Droplet

More (2013) derived a Formula for the voltage calculations from similar model analysis for his research entitled "Finite Element Analysis of Piezoelectric Cantilever" to which the author has validated the parametric evaluation of the configuration of the plate.

The Formulae for the voltage, $U_y = \dfrac{-3(d_{31})VL^2}{8H^2}$ (More,2013) (1)

Calculation of velocity, $v^2 = 2gH$ (2)

where

H=Height of free fall, considered as 3.5meters,
g=Standard gravitational acceleration viz. 9.81m/s^2)

Load developed by rain drop,

$$\text{Edge Load} = \frac{\text{K.E of the Rain Drop}}{\text{Effective surface area of the droplet impact}} \quad (3)$$

Velocity of raindrop in the 3.5m setup=8.286m/s

Figure 6. (a) The model used for simulation (b) Mesh generated on the plate

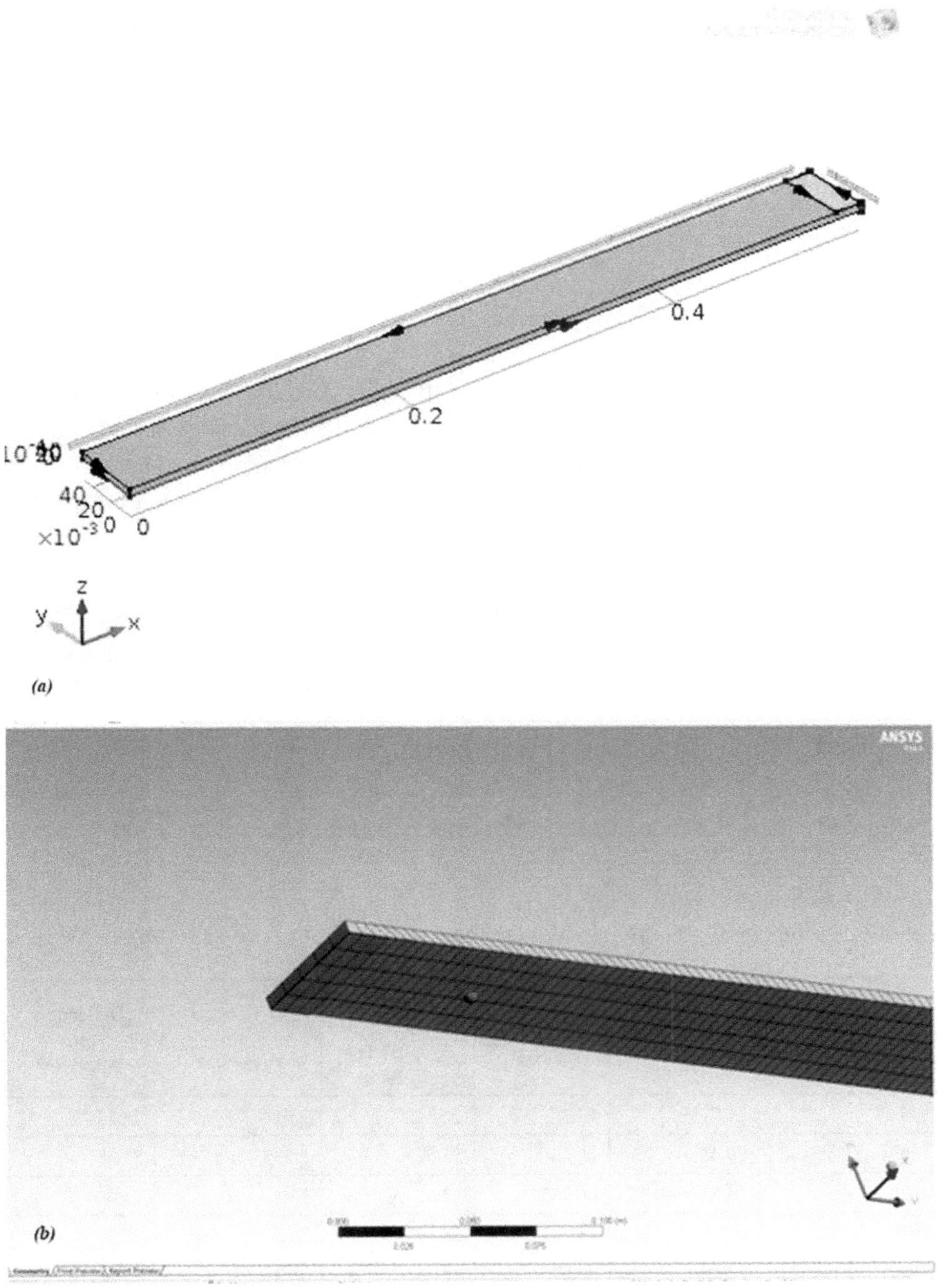

Table 3. Edge load calculation for various droplet size

Droplet size (in mm) (Assumed)	Mass of droplet(kg) $(m = \rho_{water} \times \frac{\pi D^3}{6})$	Kinetic energy(J) $(K.E = \frac{1}{2} mv^2)$	Edge Load (N/m)
1.6	2.144×10^{-6}	7.365×10^{-5}	4.419
3	1.4137×10^{-5}	4.851×10^{-4}	8.268
6	6.545×10^{-5}	2.246×10^{-3}	13.81

Validation

The author in this chapter used a 3-D Static structural simulations in ANSYS® and 3-D stationary simulations in COMSOL to conduct the simulations and to validate the results as per the parameters involved for the 2-D simulation, where one end of the beam is kept fixed and the other end is under uniform edge load (Popovici et al., 2008).

Set 1: Deflection of the Beam for Load 500 N/m

Figure 7 shows the simulation for the deflection of the beam under 500 N/m load

Set 2: Deflection of the Beam for Load 1000 N/m

Figure 8 shows the simulation for the deflection of the beam under 1000 N/m load

Table 4. Values of stress and displacement obtained from ANSYS® and COMSOL

Load (N/m)	Von Mises Stress (MPa) (Popovici et al., 2008)	Von Mises stress (MPa) (ANSYS®)	Deflection (mm) (Popovici et al., 2008)	Deflection (mm) COMSOL	Deflection (mm) ANSYS®
500	14.71	13.3	2.53	2.618	2.613
1000	29.42	26.76	5.06	5.235	5.823
2500	73.55	66.915	12.65	13.1	13.0
5000	147.1	133.83	25.30	26.2	26.1
7500	220.6	200.75	37.95	39.3	39.1
10000	294.2	267.66	50.61	52.4	52.3

Figure 7. (a) Deflection from COMSOL (b) Deflection from ANSYS®

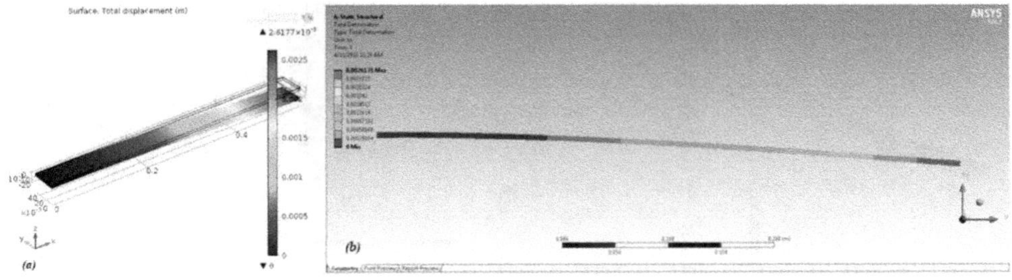

Figure 8. (a) Deflection from COMSOL (b) Deflection from ANSYS®

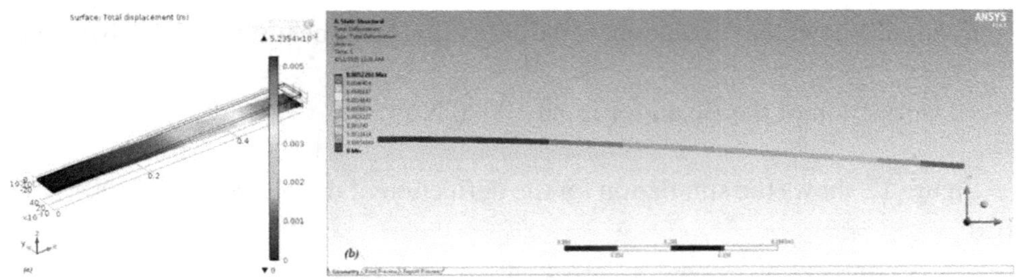

Set 3: Deflection of the Beam for Load 2500 N/m

Figure 9 shows the simulation for the deflection of the beam under 2500 N/m load

Set 4: Deflection of the Beam for Load 5000 N/m

Figure 10 shows the simulation for the deflection of the beam under 5000 N/m load

Figure 9. (a) Deflection from COMSOL (b) Deflection from ANSYS®

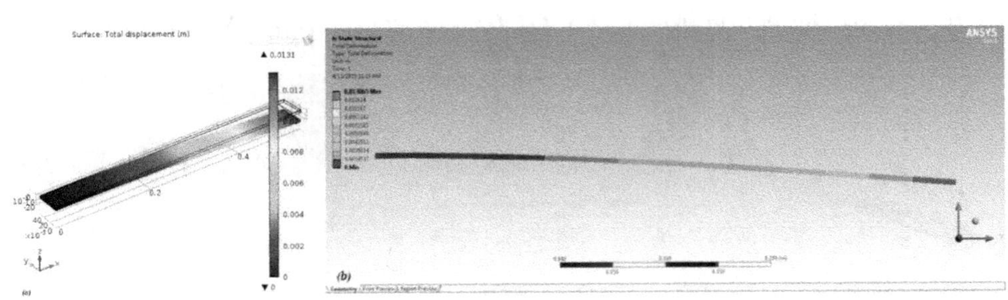

Figure 10. (a) Deflection from COMSOL (b) Deflection from ANSYS®

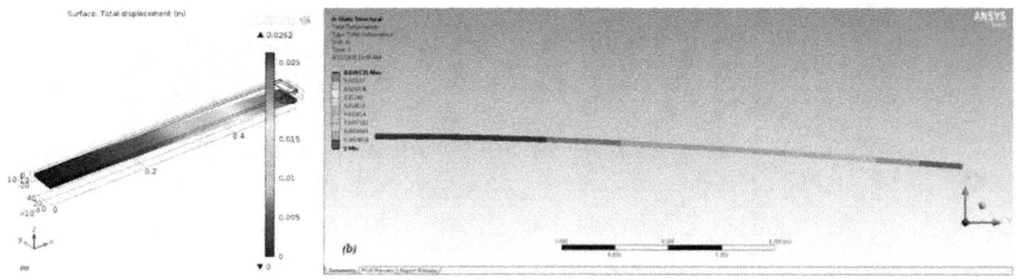

Set 5: Deflection of the Beam for Load 7500 N/m

Figure 11 shows the simulation for the deflection of the beam under 7500 N/m load

Set 6: Deflection of the Beam for Load 10000 N/m

Figure 12 shows the simulation for the deflection of the beam under 10000 N/m load

Figure 11. (a) Deflection from COMSOL (b) Deflection from ANSYS®

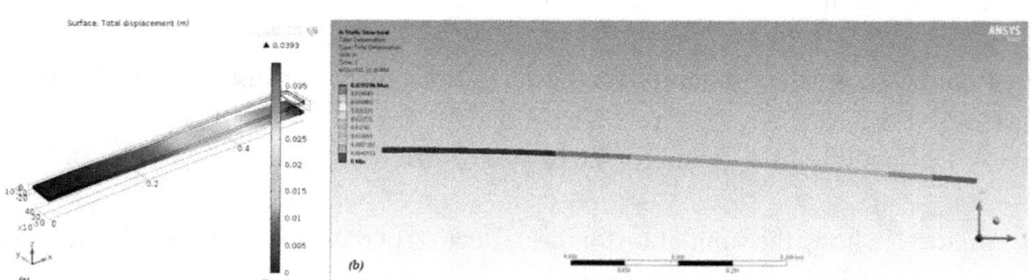

Figure 12. (a) Deflection from COMSOL (b) Deflection from ANSYS®

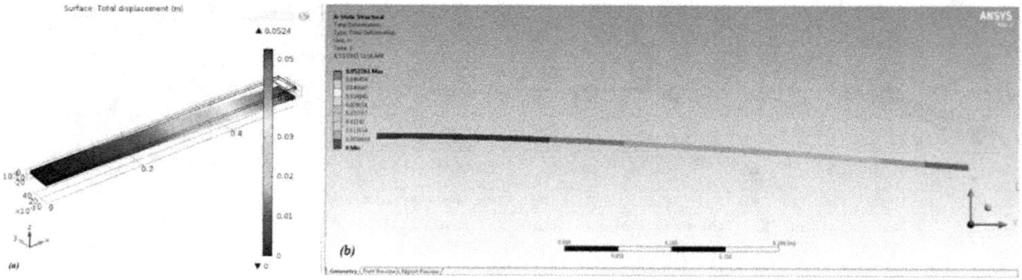

The Von Mises Stress obtained from the simulation process (ANSYS®) were then validated to the numerically obtained stress (Popovici et al.,2008).

VALIDATION PLOTS

The validations have been carried out between the variables (Load, Deflection and Von Mises stress) in MATLAB which provides a greater significance to the implicit and explicit dynamics. The author's validation in this chapter to the already existing Von Mises stress as observed by Popovici et al.(2008) corresponds the parametric evaluation to a better ratio.

Figure 13 shows a variation between Load and Von Mises Stress

Figure 14 shows a variation between Load and Deflection in ANSYS®

Figure 15 shows a variation between Load and Deflection in COMSOL

As can be seen in the above figures, the shape of the curve is almost linear to that of observed by Popovici et al. (2008). This provided a better understanding of the usage of the parameters to the concerned author in the chapter for further analysis.

Figure 13. Load vs Von Mises Stress

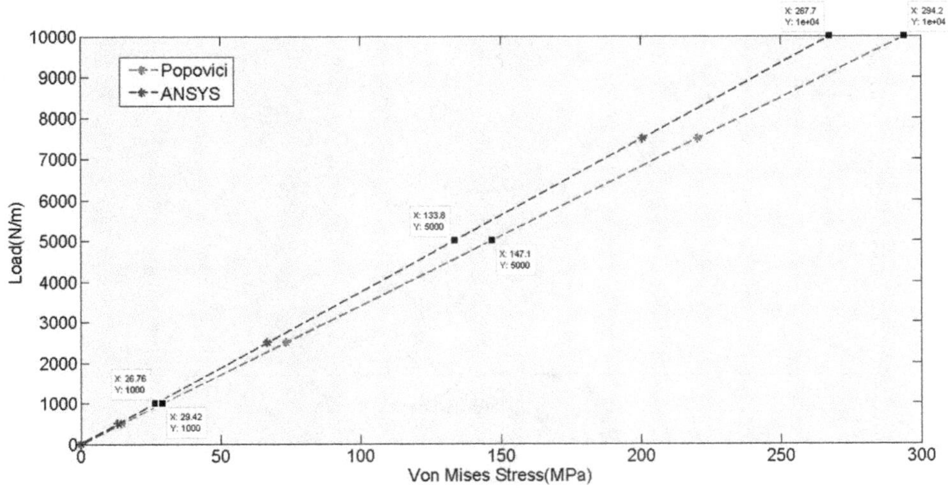

Figure 14. Load vs Deflection

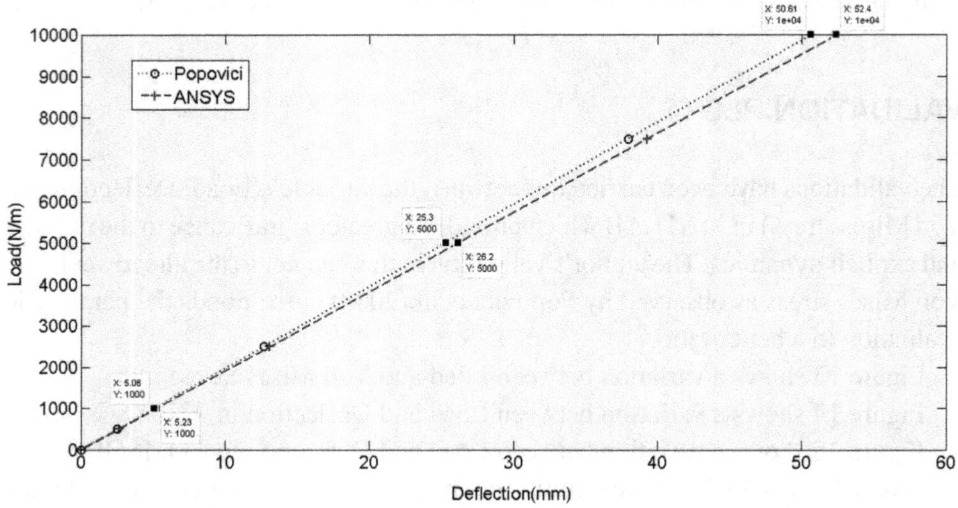

Figure 15. Load vs Deflection

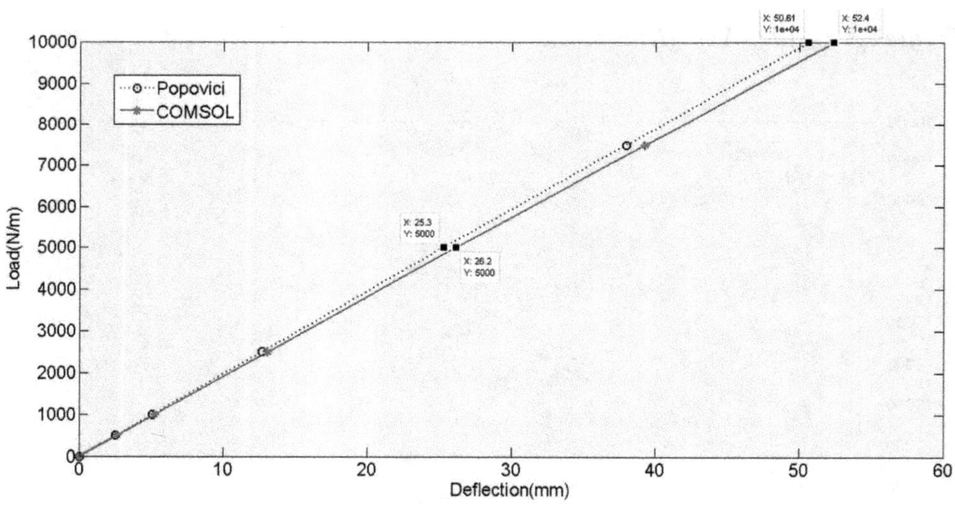

RESULTS: IMPLICIT AND EXPLICIT DYNAMICS

The validated results from the variation of stress and deflection with load created a freeway in exploring the rich diversified field of implicit and explicit study. The study dealt with the modified designs are presented in this chapter. The modified designs were prepared in steps of 1.6 mm,3mm and 5mm diameter of rain drop. With every proceeding set, the edge load varies. Corresponding to each proceeding, the deflections and voltage were obtained which served as a basis for the parametric evaluation.

However, it is seen that the multiple constraints encountered in Implicit FEM viz. static structural in ANSYS® and stationary in COMSOL are considerably negotiated by the use of Explicit dynamic study.

Implicit FEM

Figure 16 shows the Implicit results from COMSOL and ANSYS®

Figure 16. (a) Implicit result – COMSOL (b) Implicit result – ANSYS®

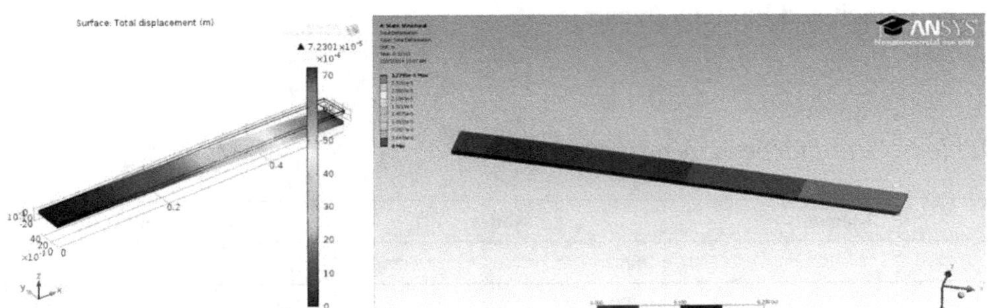

Table 5. Implicit FEM Results (COMSOL and ANSYS®) and voltage output

Rain drop diameter(mm)	Load (N/m)	Deflection of beam(in m) (COMSOL)	Deflection of beam (in m) (ANSYS®)	Voltage (in mV)	
				PZT-4A	PZT-5H
1.6	4.419	2.315×10^{-5}	1.7528×10^{-5}	0.56	0.31
3	8.268	4.3286×10^{-5}	3.2795×10^{-5}	1.28	0.48
5	13.81	7.2301×10^{-5}	5.4777×10^{-5}	1.77	0.77

While proceeding through the simulation process, whether or not the solution gets affected by the size of the grid cannot be ensured only by simple simulation study. This can only be done by the grid independence study which shows that the solution is independent of the grid size.

Table 6 gives the details of the Grid Independence Study

Figure 17 shows the variation of Number of elements with the deflection

Explicit FEM

Since, there is a static loading in the implicit FEM and the impact of the raindrop is considered on the edge. The results obtained have multiple constrains. Hence, Multibody explicit dynamic simulations are undergone, where the droplet hits the plate at 8.268m/s velocity under a standard gravitational acceleration on both the bodies.

The simulation shown in Figure 18 is carried out with a 5mm droplet size and end step time of 8.99e-3 sec.

The explicit results are shown in Figure 18

Table 7 shows the voltage output from Explicit Dynamics Study

Figure 19 and 20 shows the variation of raindrop diameter and voltage output from Implicit and Explicit Dynamics.

Table 6. Grid Independence study

Elements	Nodes	Deflection(m)
80	786	3.2607e-005
168	1492	3.2657e-005
1656	12608	3.276e-005
2200	16606	3.2767e-005
5382	34263	3.2788e-005
27500	156060	3.2794e-005
39120	220866	3.2795e-005
39600	223570	3.2795e-005
38880	219514	3.2795e-005

Figure 17. Grid Independence test

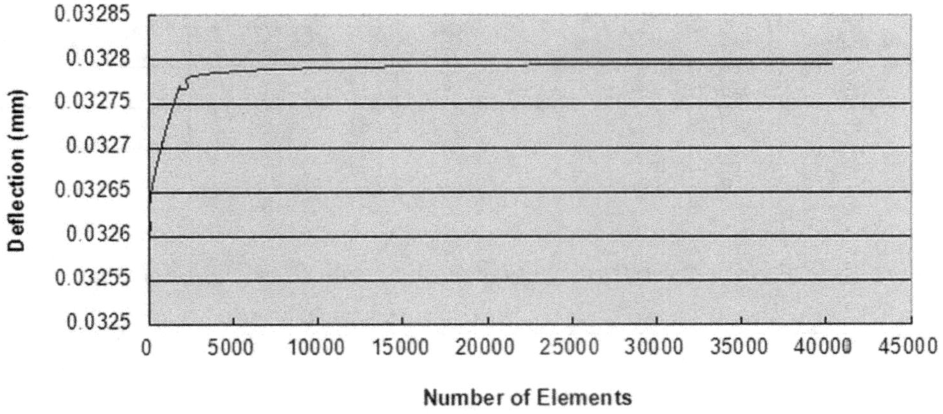

Figure 18. (a) Initial Phase of Explicit Dynamic Simulation (b) Final phase of Explicit Dynamic Simulation

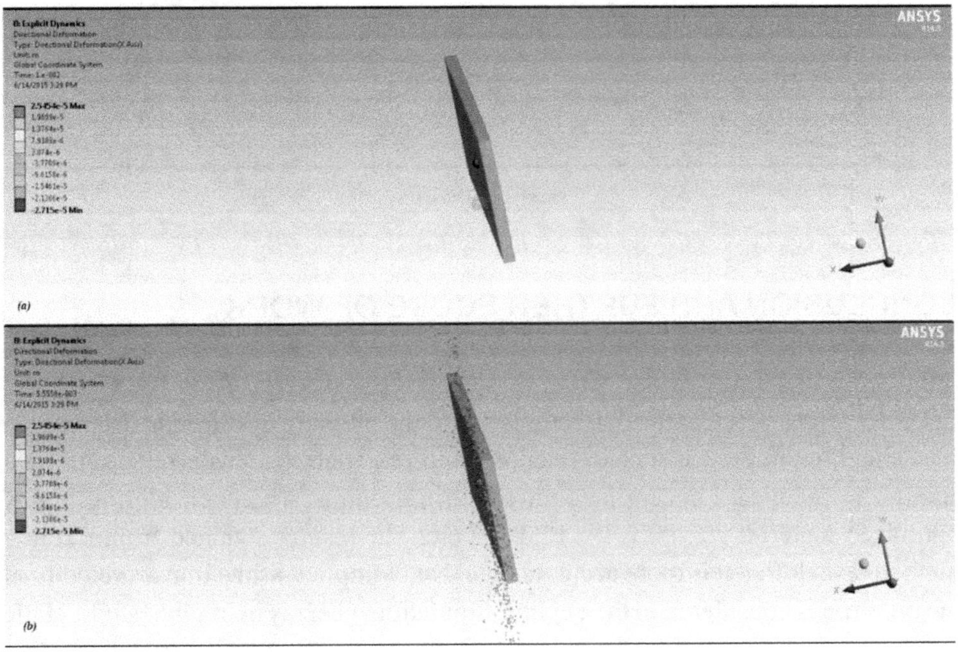

Table 7. Explicit FEM Results from explicit dynamics ANSYS® and voltage output

Drop size (in mm)	Deflection from ANSYS® (m)	Voltage(mV)	
		PZT-4A	PZT-5H
1.6	2.525×10^{-5}	1.13	.50
3	2.7×10^{-5}	1.46	.54
5	3.1073×10^{-5}	1.68	.45

Figure 19. Raindrop diameter vs voltage output graph from Implicit Dynamics

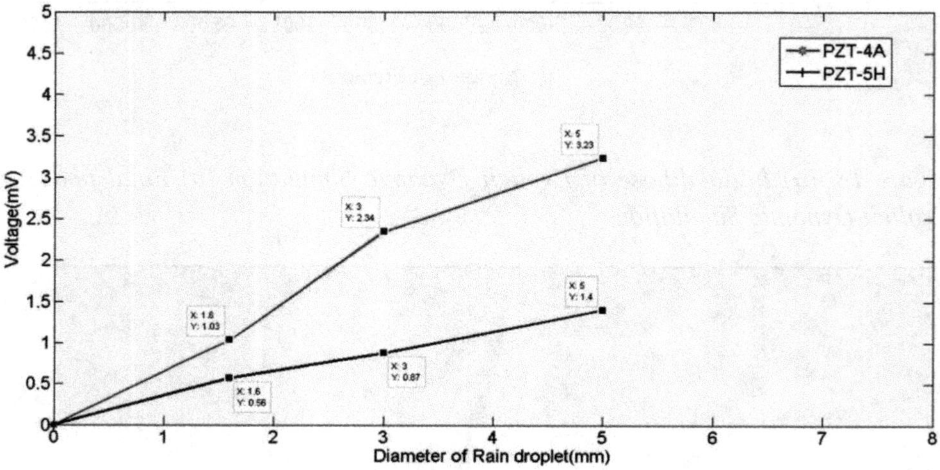

CONCLUSION AND FURTHER SCOPE OF WORK

From the initial study of deflection of piezoelectric material from ANSYS® and COMSOL, it is seen that the deflections due to the pressure by different droplets are very small but impact of such deflection on the piezoelectric material is sufficient enough to produce a decent amount of electric output when considering highly optimized material viz. PZT-4A, PZT-5H etc. The results obtained from Explicit method reveals that it is more accurate than that of implicit as the impact velocity of the raindrop is considered in the explicit simulation whereas in implicit study, static loading takes place on the edge which continues until the maximum deformation

Figure 20. Raindrop diameter vs voltage output graph from Explicit Dynamics

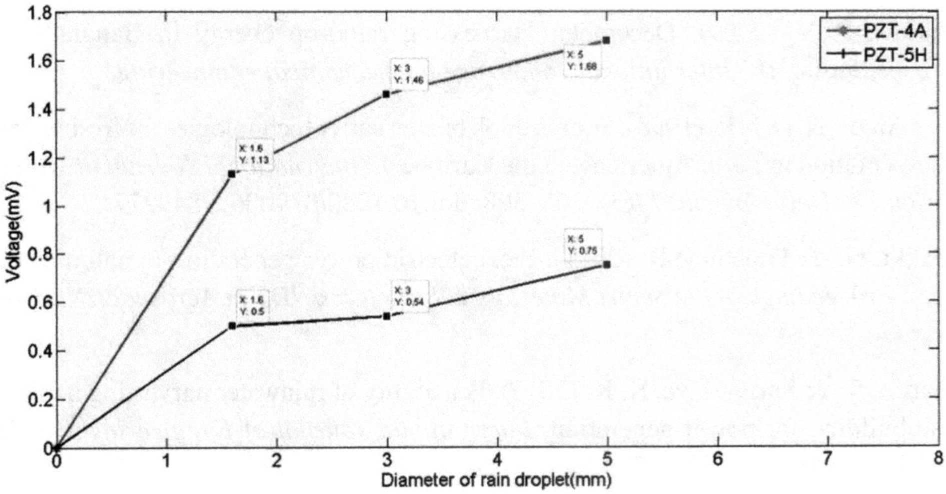

possible is reached due to the exertion of line pressure.The behaviour of material in explicit are considered as linear, hence with such negligible deflection due to the raindrop on the plate, the piezoelectric material almost behaves as an isotropic material. Hence, it can be inferred that the amount of actual voltage generated is less but with further optimization of dimensions and free-fall height of the raindrop a sufficient amount of voltage can be generated by different piezoelectric materials and the proposed method for solving can be taken into consideration for testing of different materials.

It has also been seen that the same methodology can be used to design and analyze further improvement in the performance for various combinations of input parameters. Following design modification, the parameters considered are well within safety limits. As observed above, the overall study can be practically implemented in geographical locations where the percentage of rainfall is perennial.

REFERENCES

Biswas, P. V. (2009). December. Harnessing raindrop energy in Bangladesh. *Proceedings of the International conference on mechanical engineering*.

Dos Anjos, N. D. F. R. (1998). Source book of alternative technologies for freshwater augmentation in Latin America and the Caribbean. *International Journal of Water Resources Development, 14*(3), 365–398. doi:10.1080/07900629849277

Makki, N., & Pop-Iliev, R. (2011). Piezoelectric power generation in automotive tires. *Proceedings of the Smart Materials & Structures/NDT in Aerospace/NDT in Canada*.

Martin, S., & Shrivastava, K. K. (2013). Feasibility of rainwater harvesting in high rise building for power generation. *International Journal of Engineering Trends and Technology, 4*, 522–527.

More, N. N. (2013). Finite Element Analysis of Piezoelectric Cantilever. *International Journal of Innovations in Engineering and Technology, 2*(3), 100–105.

Patel, I., Siores, E., & Shah, T. (2010). Utilisation of smart polymers and ceramic based piezoelectric materials for scavenging wasted energy. *Sensors and Actuators. A, Physical, 159*(2), 213–218. doi:10.1016/j.sna.2010.03.022

Popovici, D. (2008). *Modeling and simulation of piezoelectric devices*. In *Modelling and Simulation*. InTech. doi:10.5772/5968

Pozzi, M., & Zhu, M. (2011). Plucked piezoelectric bimorphs for knee-joint energy harvesting: Modelling and experimental validation. *Smart Materials and Structures, 20*(5), 055007. doi:10.1088/0964-1726/20/5/055007

Ramsden, Ed., & Dix, C. (2003). Low-Pass Filtering for Vibration Sensors. *Sensor Technology and Design.* Retrieved from http://archives.sensorsmag.com/articles/0203/33

Telba, A., & Ali, W. G. (2012), July. Modelling and simulation of piezoelectric energy harvesting. *Proceedings of the World Congress on Engineering* (*Vol. 2*, pp. 4-6). Academic Press.

Tinaikar, A. (2013). Harvesting Energy from Rainfall. *International Journal of Renewable and Sustainable Energy*, *2*(3), 120–132.

Wu, N., Wang, Q., & Xie, X. (2013). Wind energy harvesting with a piezoelectric harvester. *Smart Materials and Structures*, *22*(9), 095023. doi:10.1088/0964-1726/22/9/095023

Section 2

Issues Related to Investment in Different Industry Sectors

There is increasing consensus about nature providing a wide range of benefits to people and the importance of incorporating these ecosystem services into resource investment and management decisions of different industry sectors centered on sustainable and ecological development. This section explores the developments, achievements, and possibilities of specific cases related to the bioenergy sector (biodiesel companies in Mexico) and the bio-education for community development.

Chapter 5
Entrepreneurial and Institutional Analysis of Biodiesel Companies in Mexico

José G. Vargas-Hernández
University of Guadalajara, Mexico

Juan José Esparza López
University of Guadalajara, Mexico

Justyna Anna Zdunek-Wielgołaska
https://orcid.org/0000-0002-5788-9707
Warsaw University of Technology, Poland

ABSTRACT

The objective of the chapter is to analyze the role of the institutions in the biodiesel industry, in order to know if there is a relationship with the quality and maturity of the same with the ventures. Starting from a literary review, the framework of the current situation is identified, covering aspects related to formal institutions, laws, rules, regulatory bodies, and the theory that supports the relationship between institutions and entrepreneurship. The chapter concludes that the institutions in Mexico have increased their maturity and incentive to increase the number of producers and distributors of biodiesel, thus taking advantage of the growing market.

DOI: 10.4018/978-1-5225-7958-8.ch005

INTRODUCTION

The current situation of a shortage of natural resources requires the innovations and ventures in production sectors that help to reverse the situation, so the research seeks the stability of markets that achieve competitiveness and preservation of the environment.

The biodiesel market is in constant growth. By 2023 consumption is expected to reach more than 40 million liters around the world. According to OECD (2017) information, the United States, as well as European countries and some countries of South America such as Argentina, Brazil, Colombia and Chile stand out to have quickly added in this new project of an economy with a better vision of the preservation of the environment. This new change in the consumption trends of the people, generates new markets that can be exploited by developed country economies as emerging countries that have the necessary resources for the production and supply of biodiesel.

Entrepreneurship is given by several factors, some of which may be the best scenarios, such as unique opportunities, innovations that meet specific needs that were not counted in the past. These types of entrepreneurship are what the countries need because they get the most number of benefits, job creation, sustainable economic growth. The boom in the demand for biodiesel at the global level could be one of those opportunities that must be taken advantage of. However, the opportunity to recognize a growing market alone does not ensure success because the planning and knowledge of the terrain in which entrepreneurs seek to enter, are flaws that could be had. Being a relatively new market few are those who have all the necessary information to say know or be experts in the competitive bioenergy market.

For this reason, formal institutions such as laws or bodies that govern the conduct of industry can be a tool that helps entrepreneurs to have a delimited vision of the actions that are possible to carry out, which can be observed as a knowledge which is transferred from the institutions to the entrepreneurs that helps them to generate success. In addition to the existence of formal institutions with sufficient quality and maturity as can be easy to apply legal frameworks and regulatory bodies with good planning and implementation of actions, such as aid in reducing taxes. This may provide financial support and development on the part of the researchers in a specific line of research, which serves to give rise to a greater degree of entrepreneurship in a specific sector that has all the incentives.

The background check showed that there was a clear concern about the lack of rules in the game in this sector that made the market in Mexico unable to consolidate. However, there were already bodies that were looking to develop enough regulations to clarify the path of those interested in entering this new market that was emerging at a fast pace. Therefore, in this study, a theoretical analysis will be carried out, comparing the evolution of the biodiesel production companies in Mexico, in order to determine if there is a relationship with the formal institutions and they will solve the inefficiency that was seen in this sector in 2008.

BACKGROUND OF THE PROBLEM

The Mexican government promotes the production in several plants, their distribution and consumption of biofuels and among them biodiesel that promotes the use of clean and renewable energies as alternatives to fossil fuels as a response in a time when climate change has become in priority. Biodiesel is a mixture of monoalkyl esters of fatty acids that is obtained through processes such as the "transesterification" of industrial waste oils and fats. Biodiesel is a mixture of methyl esters of fatty acids that can replace diesel and is obtained from the reaction of vegetable oils or fats with methanol. Glycerin is obtained as a byproduct. Biodiesel is used mainly in Europe and the USA in mixtures with 5% or 20% diesel (B5, B20) or as pure biodiesel (B100). In Brazil and Argentina B5 is used.

In Mexico, in 2003 the Prospective of the use of bioenergy in Mexico was designed, highlighting that the infrastructure that Pemex uses to produce diesel can be used, considering that diesel engines would require minor adjustments to use pure biodiesel. In addition, it is considered that there has been an increase in the cultivation of oil producing plants in tropical and temperate climates and that there are fiscal incentives and subsidies to promote biodiesel. In 2007, a first study was conducted on the possibilities of bioethanol and biodiesel as transportation fuels in Mexico (SENER / BID / GTZ (Edit.), 2007).

The Investment Plan of the Clean Technology Fund (CTF) for Mexico was approved by the fiduciary committee in 2009. On June 18, 2009, the Regulation of the Law of the Federal Government was published in the Official Gazette of the Federation (DOF). Promotion and development of Bioenergetics (LPDB), which entrusts the Inter-Secretarial Commission of Bioenergetics to establish, review and evaluate a strategy to develop the input market ... and promote energy security, food security and sovereignty and environmental sustainability. In 2009 Biotop, which was a project for the evaluation of technical opportunities and research needs for Latin America, mentioned how the Mexican regulatory framework for the promotion of biofuels was at an early stage and was still under development. Limiting for biodiesel companies in Mexico could be due to formal institutions that did not work in the best way.

Among the major limitations of the development of production, distribution and consumption of biodiesel is the performance of formal and informal institutions of the Mexican market, mainly in aspects related to the existence of limited information, energy legislation in transition processes, structures of market, prices, costs, etc. The knowledge of this failure in the Mexican market seeks to be solved as Rembio in 2009 mentions that there were no specific promotion policies for the second generation of biofuels, but these are likely to be included in a new biofuels research program, which will be developed by The National Council and Technology (CONACYT).

The production of biodiesel is a very complex task for companies, which register a high rate of failures. To mention some companies that have failed, is the Plant in Cadereyta, Nuevo León, established in 2005 by Grupo Energéticos and the Itesm Campus Monterrey, with an annual installed capacity of 18 thousand cubic meters that produced biodiesel from animal fats and vegetable oils and that stopped operating in 2011. In Lázaro Cárdenas, Michoacán, in 2007 a plant attached to a vegetable oil factory was installed, with a capacity of 9 thousand cubic meters of annual production of biodiesel from castor oil, but its shortage caused it to close a year later. The Institute of Bioenergetics of Chiapas installed a biodiesel production plant in 2010 from Jatropha but stopped operating in 2011 due to lack of seed.

In 2013, the Secretariat of Energy developed the Mexican Atlas of Biomass, with information on the production potentials of the different types of biomass that could be used for energy purposes (Riegelhaupt, Odenthal and Janeiro, 2016).

The biomass resources used as inputs to produce biodiesel, according to their origin and technology used for processing, are divided into three generations. Riegelhaupt, Odenthal, and Janeiro (2016) develop this characterization considering that the first generation resources are fatty acids, contained in the lipids synthesized by plants and animals, mainly, the vegetable oils of the seeds and oleaginous fruits. The second-generation plants are plant lignocellulosic, residual or cultivated, whose carbohydrates can be transformed to C10 to C12 alkanes by various technological ways, such as Fischer-Tropsch synthesis, hydrolysis followed by reform or isomerization and hydrotreatment.

The third generation are cultures of microorganisms selected or modified to produce certain biodiesel precursor molecules, such as lipids and fatty acids. The current degree of uncertainty about the inputs and production processes of 3rd generation biodiesel, its costs and technological viability is very high; for this reason, there are no certain or probable values to perform the necessary calculations and to be able to formulate production scenarios and prices of 3rd generation biodiesel.

The potential for production of first generation biodiesel from oilseed crops is growing. However, the use of animal fats as an input for the production of biodiesel is technically feasible for the production of biofuels (see Figure 1).

Biodiesel is produced in Mexico with crude vegetable oils whose commercial production has not been able to be sustained due to the lack of competitive costs of the inputs or with used oils and fats that gives rise to a value chain. In this chain of value, according to Riegelhaupt, Odenthal and Janeiro (2016), several agents intervene and participate, such as the producers of the raw material that are the ones that use the oils for the kitchen, such as the traces, greaseries, processors of pork derivatives, etc., whose practices delimit the levels and quality of recovery; harvesters of the cooked oil, such as the fat collectors of butcher shops, who can contribute their efficiency in the organization for the collection to reduce the costs,

Figure 1. Cycle of production of biodiesel
(created by the author)

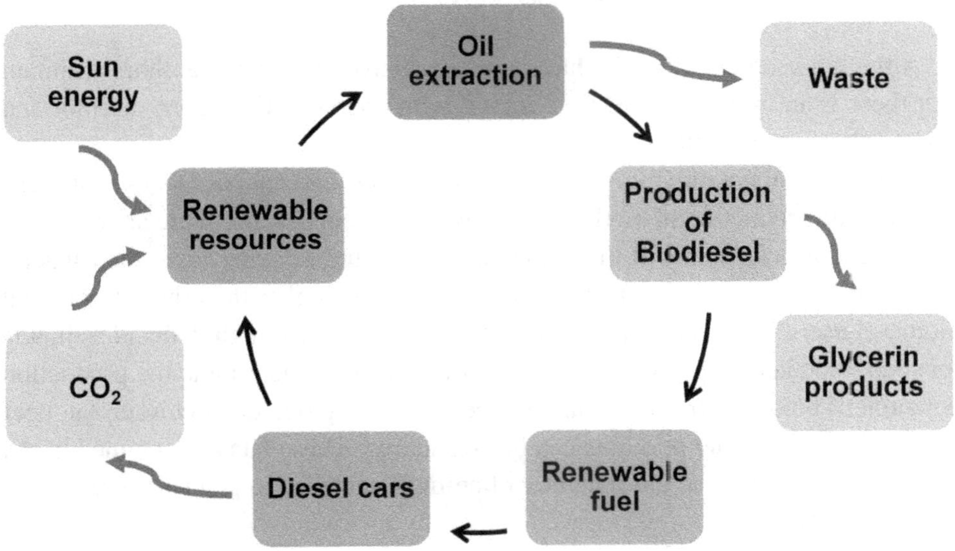

the collectors and processors that collect and refine the used oils, contribute to the conversion, distribution and sale of biodiesel and finally, the biodiesel processors who convert the oils into biodiesel, co-products and waste such as glycerin that they also distribute commercially.

Currently, in Mexico, the agents of both chains operate for the production of biodiesel, mostly informally. The information related to the capacity, type of inputs and processed volume of the companies that operate in the processing and preparation of biodiesel can be obtained from the same companies, but the information related to their marketing processes, prices and customers is unreliable and limited. Riegelhaupt, Odenthal and Janeiro, (2016) reported four companies to collect used cooking oil as the main input for the production of biodiesel:

1. Reoil Mexico collects Used Residual Cooking Oil and produces "Pre TPO" or Pre-technical plant oil that it exports to the European Union where it is processed in two biodiesel plants.
2. MORECO collects used cooking oil and transforms it into biodiesel in three plants installed in Michoacán, Guanajuato and Querétaro.
3. Biofuels de México collects used cooking oil and produces biodiesel in Puebla, Toluca, Cuernavaca, Tlaquepaque, Zapopan, Tonalá, Veracruz, Cordoba, Xalapa, Boca del Rio, Queretaro, Cancun, Playa del Carmen, Chetumal.
4. SOLBEN, from Monterrey, Nuevo León, collects used cooking oil, produces biodiesel, sells biodiesel plants and provides services such as characterization of seeds and oils, technical support, automation and biodiesel quality analysis.

Although used cooking oils have few alternative uses, some authors maintain that there is an informal collection circuit, with operators that filter, discolor and package the recycled oil to sell it as edible oil.

The technical potential for obtaining cooking oils and animal fats to be used as inputs in the production of biodiesel, is only theoretical because there are currently concurrent uses for many of these sources of fats, and in many cases the capacity and willingness to pay of concurrent users would be higher than that of potential biodiesel users. This is the case of animal fats, which have high demand as inputs for soap, food additives, pet food and other end uses. To determine the production potential of biodiesel from animal fats, the use, among others, of chicken fat, beef tallow (bovine fat) and pork fats can be considered. These fats are obtained in the traces, as co-products of the slaughter of animals, and also in plants that reprocess butcher's fats.

Based on official statistics of SAGARPA (2017, 2010), obtained through consultations with the Agrifood and Fisheries Information Service (SIAP) to determine the potential for biodiesel production from current oilseed crops. they are considered to be the most important sources of fatty acids at the global. Seed to oil conversion factors are an average of industrial performance for pressure extraction technology. For soybean and sunflower, two sequential extractions are carried out, the first by pressure and the second by organic solvent. Riegelhaupt, Odenthal and Janeiro (2016) analyze that for the case of those considered in group 1, as new crops, it is integrated by the Mexican pinion (Jatropha curcas) and the castor bean (Ricinus communis). Their contributions to the national production of oils are minimal, and their trends are still undefined.

The minor crops considered in group 2, is composed of canola (Brassica rapa) and sunflower (Heliantus annuus). The canola has low yields, but the sunflower has experienced a significant increase in its cultivation and found a market niche of better prices as a specialty. In the third group, considered as dynamic crops, it is integrated by coconut (Cocos nucifera), oil palm (Elaeis guineensis) and soybean (Glycine max), crops that have had a clear tendency to increase the harvested area and / or the production.

The potential of agricultural residues, esquilmos and stubbles considered as the parts of the vegetables, usually stems and leaves that are not removed from the field during the harvest is estimated at 15 million tMS / year. Riegelhaupt, Odenthal and Janeiro, (2016) assume a conversion rate of 0.25 m3 of biodiesel / tMS based on the average of the values reported for pilot plants and plants projected according to the European Biofuels Technology Platform (EBTP, 2016). Assume a technically sustainable production level of about 3.75 million m3 of biodiesel per year. However, the authors consider that this potential is not economically feasible for the production of biodiesel in the short and medium term, since the conversion technologies known as 2nd generation have not yet reached the degree of technical maturity to be applied on a commercial scale.

The largest potential source of woody forest biomass in Mexico as a raw material for second-generation biodiesel is the native forest vegetation. The estimate of sustainable annual production excludes wood of species and qualities that can be used by the sawmill, cellulose and panel industries, and includes only wood that is useful for energy. The total potential national production calculated by Riegelhaupt, Odenthal and Janeiro (2016), counts forests and jungles located less than 3 km from a road, with land slopes of less than 30%, outside protected natural areas and

/ or areas of high value for the conservation of biodiversity, is 112 million tMS / a, but some 22 million tMS / a currently used as domestic fuel and 3 million tMS / a destined to manufacture charcoal should be deducted.

The net potential is thus about 87 million tMS / a. This potential is found in almost all the states of the country, and its annual harvestable increase ranges from 2.0 to 4.2 tMS / ha / a. It is therefore similar to the harvesting rates of field residues or agricultural residues, but it has the advantage that the growth can be accumulated in cycles of 10 to 15 years, to harvest from 30 to 50 tMS / ha in an operation at the end of each cycle. As with agricultural wastes, Riegelhaupt, Odenthal and Janeiro (2016) argue that the conversion of woody biomass into biodiesel is technically possible, but there are no economically viable industrial processes so far.

For the development of future scenarios for the introduction of biodiesel in Mexico, several elements have to be analyzed along the biofuel value chain, from the production of inputs to the commercialization of the fuel market and including the conversion infrastructure of biofuels. inputs to biodiesel. The level of future penetration of biodiesel in the Mexican fuel market depends on a large number of economic, technological and institutional factors, as well as political decisions. It has been noted in the empirical review that a problem in Mexico for entrepreneurs interested in the renewable energy markets in specific biodiesel is that the institutions are not clear enough in their actions that helps to understand the market.

The SMEs producers of biofuels to survive in the market have to fight against their main competition, the PEMEX emporium, producer and marketer of fossil fuels. However, the excessive increase in the price of gasoline, alternative fuels, such as biodisel, are becoming more competitive in the market. However, the biggest challenges for SMEs producing biodiesel are marketing processes.

To analyze the demand of the fuel market, the price that buyers would be willing to pay for biodiesel is determined. Biodiesel is a substitute for diesel of fossil origin, and for this reason the existence of demand for biodiesel depends mainly on the reference price of diesel of fossil origin defined as the price before the application of taxes. If biodiesel can be produced (or imported) and marketed at prices lower than the diesel reference price, there will be demand for biodiesel in the market. If, on the other hand, the reference prices of diesel of fossil origin are lower than those of biodiesel, there will be no demand for alternative renewable fuel. If there is no demand for biodiesel in the market, this can be induced through economic and fiscal support policies to achieve a certain level of penetration of biodiesel in the market or, alternatively, determine the market share that can be obtained and sustain with a certain level of spending on support policies.

There is uncertainty about the supply of bioenergy and biofuels based on renewable resources in the market, as a consequence of the lower price of fossil fuels. This situation means that there are few economic incentives to invest in alternative fuel technologies. To make the production, distribution and consumption of biofuels more efficient and cheaper, institutional aspects such as the legal framework and the supply chain must be resolved. If reached by 2030, the greatest impact will be on the capacity to generate a significant reduction in carbon emissions.

The existence of demand for biodiesel in the fuel market is complemented by the evaluation of the evolution of the installed capacity of infrastructure for the conversion of inputs to biodiesel, to determine that it is sufficient to satisfy the demand at all times. Finally, if there is both demand for biodiesel in the market and conversion capacity, it is required to evaluate the availability of inputs as well as their production costs, or import to produce the biodiesel demanded in the market.

DELIMITATION OF THE PROBLEM

Given the growing market for biodiesel consumption and the fact that there are no companies that monopolize production, a solid production base can be created in Mexico to help the country's economic growth. However, no solid evidence has been found to be advanced at the pace that should be.

Considering that natural factors are conducive to the production of the raw material and there is no monopolizing company of production in Mexico, adding that the demand for biodiesel is increasing, it would be expected that by complying with the theory of resources and capabilities and that the industry is not strong, the biodiesel production sector is increasing. This leads to conclude that there is some other factor inhibiting this growth, and the antecedents of the lack of institutions that help the initiatives of productive companies in biodiesel generates the following question:

How have formal institutions impacted the creation of biodiesel producing companies in Mexico?

JUSTIFICATION

As the biodiesel sector is a growing market that demands more liters' year after year, it is an opportunity for Mexico, to consolidate its productive industry and to be competitive worldwide.

From 2014 consumption has approached the production with its increase of 1454.6 per year in average consumption compared to the 1436 that increases per year on average the production. It is concluded that, if this trend continues, the demand to consume will be higher than the production, which is why it is a moment for there to be ventures in this sector. The estimators of the variations in the independent variable are good estimators because the coefficient of determination (r^2) is greater than 6 which is a parameter that is used to accept (Tapia 2017).

However, there is concern that institutions have been able to improve the picture, so that enterprises in this sector have a better direction. The Energy Regulatory Commission in (2014) mentions the regulatory area in which it is in charge of energy, in addition to mentioning the institutions that also help regulate it, such as the National Hydrocarbons Commission (CNH), the Secretary of Energy SENER), the Ministry of Finance and Public Credit (SHCP), the Federal Economic Competition Commission (COFECE) and the Security, Energy and Environment Agency (ASEA).

The objective of the institutions is to create conditions for the efficiency of the energy markets (Energy Regulatory Commission 2015). There is a growth in the regulatory framework of bioenergetics, for which it is necessary to know if the stability in the regulatory framework that has been established has helped in the stability of the market and above all to know if the institutions have favored that there are ventures in biodiesel.

There is a relationship with institutions and entrepreneurs around the world as research in 2015 mentions that

To the extent that the entrepreneur carries out his activity in an environment where the rules of the game are well-defined (i.e. that there is good protection of property rights, that the judicial system functions properly, that governments are stable or that there is no corruption, among other factors), the entrepreneur will enjoy greater security that will positively contribute to the success of his company (Fuentelzas and Gonzalez, 2015).

THEORETICAL ASSUMPTION

Institutions understood as the limitations to the behavior of an individual or a society, have a very important role in the creation and consolidation of MSME's biodiesel companies concerned. Any entrepreneur or entrepreneur must abide by them for the proper functioning of their organization. Leaving the institutional framework entails tangible and intangible sanctions for the organization. First, the institution is defined as the humanly constructed constraints that structure human interaction (North, 1990). On the other hand, the institutional framework is described by Peng (2012) as the set of formal and informal organizations, which are responsible for regulating the human behavior of people and companies. These are based on three pillars, which are: regulators, normative and cognitive.

As mentioned above, institutions are divided into both formal institutions and informal institutions. Within formal institutions are found the laws, regulations and rules to which every individual who interacts in a society must be subject or otherwise have sanctions, as described in the regulations and laws. As for formal institutions, all MSME's that are established in Mexico will be subject to the Federal Law of mercantile companies. This law is in charge of regulating anonymous companies or not, codes or state laws and municipal regulations on the opening of establishments, which depend on the geographical location where the MSME's biodiesel company is established, as it will change depending on the state or municipality: Income Tax Law (ISR), Single Tax Rate Business Act (IETU), Value Added Tax (VAT), Industrial Property Law, Federal Data Protection Act, Federal Labor Law, Social Security, Law of the Institute of the National Housing Fund for workers. Each law has a particular objective which seeks the regulation of all companies in Mexico.

On the other, side of the institutional framework are informal institutions which include norms, culture and ethics that apply. According to Peng (2012) these formal institutions have two supports: normative and cognitive. The first refers to how it influences the behavior of people and in this case companies, the values, beliefs and actions of relevant actors. On the one hand, of the cognitive pillar, refers to the internal values and beliefs that guide the behavior of individuals and companies. Some of the informal institutions are covered by the culture, beliefs and values of a society and that in some way affects the functioning of MSME's in Mexico.

Culture can be defined as the collective programming of the mind of individuals. This programming or way of thinking distinguishes the members of one group or category of people from another (Hofstede, 2011). Ethics is also part of all informal institutions, defined as "Norms, principles or standards of conduct that govern the individual behavior of organizations" (Vargas, et al. 2005, p.133). Kaushik Basu explains the importance of social norms and culture in the functioning of the economic system, in the first instance it should be understood that all human beings are socially designed to behave in a certain way respecting the social norms that exist (Basu, 2013). All the social norms in which action delimit our behavior and, in this case, delimit the action of the MSME's, each regulation or law limits the possible actions that the companies can carry out.

The central question is What do institutions do? Or why are they so necessary? According to Peng (2012) the role of institutions is to reduce uncertainty because they indicate the legitimate and illegitimate channel, what institutions do is to delimit actions to only acceptable leaving out those outside this range and thus reduce transaction costs. Reducing uncertainty leads to a reduction in transaction costs which are vital for the proper functioning of any economy. The less transaction costs, the economy of a country or a locality function more effectively. Transaction costs were introduced by Coase and defined as the costs of market use (Salgado, 2003). Transaction costs are only the costs of information search, negotiation costs, decision making, formulation costs, and policy implementation. Another definition for transaction costs is the one proposed by Peng (2012) which defines this concept as: the costs associated with economic transactions, or the costs of conducting negotiations.

Institutions, both formal and informal, play a vital role in the reduction of transaction costs, without the existence of these the operation of each transaction in the economy would be much greater. In developed economies transaction costs tend to zero, while in developing economies these costs are much higher. This is explained by the weakness of the institutions, which makes the uncertainty high, which is why it is necessary the signature of contracts between individuals.

In the framework of informal institutions according to Peng (2012) these institutions reduce uncertainty in various ways. The first is related to relationship contracts, known as "personalized exchanges", informal and based mainly on relationships. Robert Putnam's theory of social capital brings much to the subject of reducing transaction costs through informal institutions. Social capital is understood as "the characteristics of social organization, such as networks, norms and trust, which facilitates coordination and cooperation for mutual benefit" (Urteaga, 2013, page 45). These networks of confidence through expected behavior cause confidence among economic agents to be high so that transaction costs will be null because expected behavior.

One concept that helps to understand the importance of informal institutions is "The Culture of Confidence: one can trust the given word because that is the social norm, so people follow it by instinct" (Basu, 2013, p. 167). The main reason why the culture of trust is important is because it is needed to rely on the other's word on multiple occasions it is not feasible to sign any contract at any time and then go to court in if the other party does not comply with the agreement, it is not feasible because of the high costs that the operation will have. That is why the culture of trust between two parties is important.

Another important concept within the theoretical review of formal and informal institutions is institutional transactions, which are defined as major changes that are incorporated into the formal and informal rules of the game that affect organizations as players (Peng, 2012). Property rights are another of the formal institutions that influence business. These rights are social institutions that define the range of privileges granted individuals to specific resources. In other words, it is the right of individuals to direct or indirect power on an object or a good (Mahoney, 2010). The theory of property rights was developed by various theorists including Ronald Cease to explain the social and economic relationships that people should consider when using scarce resources.

From this discussion, it can be developed the following assumption:

Formal institutions positively impact the entrepreneurship of biodiesel companies in Mexico.

THEORETICAL CONCEPTUAL FRAMEWORK

Emergence of Biodiesel

With the emergence of the industrial revolution by force, the great population growth around the world arises some problems, as Agarwal (2007) mentions humanity faces problems, in which arises the main interest on the environmental deterioration and the crisis of the industry energy. One of the main causes of air pollution is the burning of oil, since the combustion of air produces large amounts of greenhouse gases (carbon dioxide, oxides of nitrogen), sulfur oxides, unburned hydrocarbons and ash thin (Agarwal, 2007). In addition, it needs to be considered that hydrocarbon-based fuel is a non-renewable good, it is finite on the planet and therefore it has been necessary to resort to energies that can be renewable and more friendly to the environment.

Recent studies indicate that there are other sources of energy, which have extremely low emissions and appear to have the potential to become energy sources for automotive propulsion, including alcohols, natural gas, hydrogen and biodiesel (Medina, Chávez and Jáuregui, 2012, 63).

This leads to the conclusion that an alternative for this destruction that causes economic growth inconsiderate by the environment, may be the use of these substitutes for conventional fuels (Figure 2).

Figure 2. Projection of biodiesel production and consumption from 2013 to 20123 (created by the author on the basis of data extracted on 25 Jan 2019 08:46 UTC (GMT) from OECD. Stat (OECD Stats, 2017))

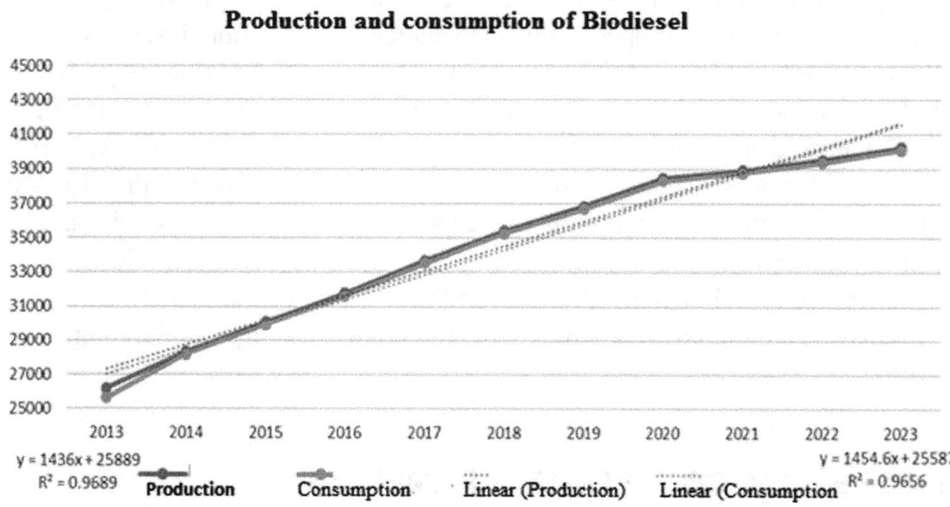

The definition of biodiesel used for this work is as follows:

It is a mixture of methyl esters of fatty acids (FAMEs) that is obtained by a process of transesterification, mainly from vegetable oils and animal fats and, more recently from algae. It can be used as a substitute for conventional diesel, as it has similar properties, reducing pollution and extending the life of the vehicle's engine. In its production, glycerin is obtained as a by-product, with multiple industrial applications (Molina, 2012).

Theories of Entrepreneurship

First, it is necessary to state that it is what is understood as an entrepreneurship. The entrepreneurship is the search for constant change, generation of ideas, speculation of projects that generate beneficial effects for the economy and society (Contreras and González, 2010). In a broader explanation, Sánchez and Pérez (2015) define it as those initiatives of an entrepreneurial nature that relate the way of being an individual where are included the way of thinking, reasoning and acting, with the search for a business opportunity, which give as a result initiative that generate value not only in an economic sense. In addition, concerned with social aspects, where all the actors are involved in the initiative, which go from owners, employees, customers, suppliers and society in general that enjoy this benefit.

However, Baumol (1990) makes a distinction between undertakings that are productive, unproductive and destructive. The productive enterprise is one in which there really is innovation, economic growth and can be associated with the discovery of good business opportunities. There are more distinctions between what are entrepreneurships like the one that is done between entrepreneurship by opportunity and entrepreneurship by necessity. When there is an opportunity venture, there is greater growth in the economy because that innovation has found a new niche market, which is exploitable. What does not happen when individuals are forced to undertake by necessity, that the development of jobs and growth of the economy is not with the same acceleration (Reynolds, Bygrave, Autio, Cox and Hay, 2003).

The World Economic Forum also continues to classify high-impact entrepreneurship as one that generates greater benefits for consumers through the use of innovations, which in turn also generate more job opportunities and greater wealth for owners (Economic Forum World, 2014).

For all in their classifications they approach the same point in which they describe the entrepreneur as the one who manages to find opportunities that generate greater benefits and the use of innovations that sustain and exploit the opportunity more time. These entrepreneurs who focus on their clients are the ones who generate the greatest wealth for their owners (Hitt, Ireland, Sirmon and Trahms, 2011). These entrepreneurs tend to plan their actions better and to enter the market early and the most usual is that they undertake in markets with which they have already had interaction, which makes their rate of failure less than in situations in which there is no such planning and market knowledge (Liñán, Fernández, Romero, 2013).

Entrepreneurship is an indispensable mechanism for the development of economies (Schumpeter, 1934), it is necessary for every economy to develop enterprises that foster the generation of valuable jobs and innovations. However, the view was that political rulers should focus on favoring the growth of developed firms with the greatest potential for growth, leaving behind small firms and ventures (Friar and Meyer 2003). The political rulers who are responsible for the development of public policies, which Bazua and Valenti (1993) explains that a public policy is a reference to any social decision, which can be taken by an individual, an organization or the state.

Governments at all levels from federal, state, and local to the importance of firms on the economy, social and political issues have focused on creating the policies needed to support entrepreneurship. But they are faced with the decision whether policies will focus on supporting the nation's leading firms, helping existing firms achieve survival, or focusing on the formation of new enterprises (Stam, Studdle, Hessels and Stel, 2009).

Theories of Formal and Informal Institutions

North (1990) defines institutions as the humanly constructed constraints that structure human interaction, which is understood as the "rules of the game". Companies must take into account the terrain in which they will enter to know how they can act and if it is possible to take advantage. In addition to mentioning that institutions are created to reduce uncertainty. These institutions that are created to reduce uncertainty can encourage investment to increase productivity, but in countries where institutions are not as robust, the effect is the opposite (North 1990).

Peng (2012) states that the institutional framework is constituted by both formal and informal organizations, which will mark the actions that can be taken individually and collectively. Formal organizations are institutions that are represented in writing and are usually more rigorous in making them comply with the desired pattern of behavior; These are the laws, the regulations and the rules. However, for the study of this research, the informal institutions that although in many investigations show that it is one of the main factors to consider when entering new markets of which knowledge is lacking, this study only focuses on analyzing how the formal institutions favor or not the ventures.

Undoubtedly, the levels of entrepreneurship in a given country or region are conditioned by the environment in which entrepreneurship is carried out, since such an environment can facilitate or discourage the initiation of new initiatives (Fuentelsaz, Gonzãlez and Macías 2015). Institutions of a formal nature, such as laws or regulations, make it easier to visualize how processes should be carried out, which makes the business easier to manage and if it does not have experience, as is the case of ventures having a well-defined way of acting can achieve success. Proprietary rights, commercial legislation, constitution procedures, ideas, cultural beliefs, gender, attitudes towards the entrepreneur, etc., influence the appearance and development of new companies (Marulanda and Morales, 2016, 18).

Based on the theory of institutions, Urbano and Diaz (2009) affirm that the environment in each country will be decisive in terms of business opportunities that are available in addition to the perception about them and their ability or capability to adaptation to use them in their favor will generate greater incentives for the creation of new companies.

Without the necessary experience, in the new environment in which new start-ups seek to penetrate, they may encounter problems that limit access to financing, make it difficult to hire high-quality personnel and result in higher transaction costs

(Aldrich and Auster, 1986). Under the circumstances of uncertainty on the part of those seeking to undertake an initiative, Fuentelzas (2015) mentions how the existence of solid institutions can facilitate the resolution of complications of inexperience, access to resources necessary for its functioning, and stability in the development of its activities. Institutions can facilitate access to resources and provide the necessary stability for the development of the activity, with the consequent effect on levels of business survival.

This demonstrates that the quality of institutions has a strong influence on competitiveness and growth, affects investment decisions, the organization of production and plays a key role in the way societies distribute profits and the costs of developing strategies and policies. However, an unstable institutional environment leads to even more complicated behavior among firms, hinders their cooperation and makes the vision of growth in the short term, which has negative effects on the quality and success of firms. (Fuentelzas and González, 2015).

The uncertainty faced by the initiatives means that the number of successful ventures is reduced, however, as Vargas, Guerra, Bojórquez and Bojórquez (2014) mention that institutions have as their main role to reduce uncertainty and distinguish between uncertainty in a) Policy, which refers to ethnic disturbances that can cause problems for firms, and b) economic, which are all transaction costs that can be given by opportunism, to prevent these negative circumstances in the firm refer that firms can make use of contracts.

CONTEXTUAL FRAMEWORK

The main regulatory framework in Mexico is the political constitution of the United Mexican States that establishes the framework of action of the authorities and the governed, through principles, rights and obligations that govern the rest of the national legal framework. Article 4 of the Constitution speaks of "Everyone has the right to a healthy environment for their development and well-being". "The damage and environmental deterioration will generate responsibility for who provokes it in terms of the provisions of the law" which has been added as Decree of the Official Gazette of the Federation on February 08, 2012. In article 25, the constitution mentions that "the State is responsible for national development to ensure that it is comprehensive and sustainable." This demonstrates its concerns about caring for the environment.

The main law in specific terms in renewable energy is the Law on the promotion and development of bioenergetics that in its first article mentions "to promote the production of inputs for bioenergetics from agricultural activities, forestry, algae, biotechnological processes and sovereignty food "its main obligation under this law is to promote the use of bio energies in Mexico. Article 8 of this same law mentions that "the Bioenergetics Committee is created, which will be composed of the members of SAGARPA, SENER, SEMARNAT, the Secretary of Economy and the Secretary of Finance and Public Credit" with which several organizations are involved in the promotion of bioenergy production and research.

Likewise, the energy agency in Mexico, the Secretary of Energy (2017), reports that 7 permits are granted for the production of bioenergetics, 41 bioenergy marketing permits, 5 transport permits and 21 permit exemption notices for Production of bioenergetics.

The National Commission for Science and Technology in February 2017 reports that the Secretary of Energy launches a call for the creation of development projects in sustainable energy technologies that comes with the support of the Bank of Mexico in which the first prizes will be awarded prizes in cash of up to 2 million dollars for the first place, from 500 thousand dollars for the second place and 100 thousand dollars for the third place. There are these types of competitions that encourages the research and development of initiatives in alternative energy with which firms who want to venture but lack financing. This type of competitions can be the solution, these are not isolated situations, it is a way to encourage development that has its opportunity year after year.

The Ministry of Agriculture, Livestock, Rural Development, Fisheries and Food (SAGARPA) (2016) reports that it has promoted from 2013 to date 960 projects for the production of biofuels in Mexico. In the last four years' incentives have been granted around 275 million pesos, which has generated a total investment of 529 million pesos. And on the installation of 7 biofuel plants, six of which are dedicated for the production of biodiesel with 4 plants more than they existed in 2009.

METHOD

This article corresponds to a qualitative research based on a review of the literature on studies applied in Mexico about the characteristics, factors and variables that Mexican biodiesel companies must possess as well as data presented by institutional organisms and comparing with the applied theories of entrepreneurship and institutionalism to appropriate innovation and new forms of administration to the needs of 21st century companies.

The nature of the research is exploratory and descriptive through the analysis of previous research trying to provide quality and flexibility that allows researchers to use multiple data collection methods with the aim to analyze the role of the institutions in the biodiesel industry, to explain and describe the phenomenon under study and to determine if there is a relationship with the quality and maturity of the same with the entrepreneurial ventures.

ANALYSIS OF RESULTS

An attractive market such as biodiesel for its constant growth should have a sufficiently strong regulation to encourage the consolidation of existing companies as well as new entrepreneurs seeking to enter into competition for the market, so that it was imminent that the State should intervene to promote economic growth. The lack of competitiveness in the sale prices of biodiesel with the reference price marked by diesel fuel of fossil origin, the demand for biodiesel can be induced through different formal institutions and support policies.

For example, the analysis of Riegelhaupt, Odenthal and Janeiro (2016) point out, among other options, the establishment of a legal obligation in the form of a quota to certain actors in the market, generally fuel trading companies, to guarantee that a fraction of the total consumption of diesel is replaced with biodiesel. There are different possible designs for the quotas depending on who are the obligated parties, the obligation to mix in the final product or if different mixtures of fossil diesel-diesel are allowed as long as the quota is met globally. An important element in the design of the obligation is the price of the penalty established in case of default to the obligated parties. The price per liter of fuel that the obligated party must pay in case it cannot provide evidence of having marketed the volume established in the obligation. The establishment of a quota guarantees the achievement of a certain penetration of biodiesel in the market. However, the legal obligation by itself does not control the unit cost of support needed per liter of biodiesel introduced in the market,

Another support policy option proposed in the analysis of Riegelhaupt, Odenthal and Janeiro (2016) - which can be implemented independently or in combination with a quota - is to provide economic incentives to improve the competitiveness of biodiesel in the market. In Mexico there are no recent statistics on the production and use of biodiesel, but it is expected to increase due in part to the increase in the price of fossil diesel. Although fiscal incentives would be needed, that was what triggered the development of the biodiesel market in other countries.

The mechanism can be the reduction of production costs (for example through subsidies or tax exemptions to farmers or producers of biodiesel) or the reduction of final marketing prices through monetary support or tax reductions. Whether the economic incentive is established as a subsidy to biodiesel producers or if it is established as a tax exemption, the result sought in both cases is to increase the competitiveness of biodiesel in the market. In this study we refer to this type of political instruments as "economic incentives". The establishment of an economic incentive (e.g. a tax reduction) allows in principle to control the level of support per liter of biodiesel, but does not guarantee by itself the control of the volume of biodiesel introduced in the market.

The two support options described result in additional direct costs to society. These are calculated as the differential between the cost of introducing a certain amount of biodiesel in the market versus that of continuing to use fossil diesel. However, both instruments differ from the point of view of the distribution of said costs among different actors. The establishment of the mandatory quota generally transfers the additional cost of acquisition to suppliers, who in turn transfer the cost to the final consumer. In the case of economic incentives, these usually leave the state budget in the form of tax reductions, tax exemptions or subsidies. To meet a certain demand for biodiesel in the market, not only the necessary inputs must be available, but also sufficient production capacity in the supply chain. A key element of the supply chain is the facilities for converting inputs to biodiesel.

The production of biodiesel requires institutions to consolidate the creation of an agency responsible for regulating and facilitating the development of the biodiesel industry in Mexico. Small producers of biodiesel such as Smart Recycling of Puebla, has the permits from the Ministry of Energy for the production, sale and transportation of biofuel and certification processes for the quality of biofuel, values such as viscosity, density and pH, parts per million glycerin and some waste.

Due to the risks to sustainability that biofuel production could present if not done correctly, several initiatives have been developed to certify liquid biofuels produced in a sustainable manner; among them the RSB3 and the GBEP4, which establish principles and criteria such as: conservation of biodiversity, non-deforestation, GHG mitigation, good use of water resources, respect for human and worker rights, among others. In Mexico, the Ministry of Environment and Natural Resources (SEMARNAT) is developing the voluntary NMX standard for the certification of the production of liquid biofuels of plant origin based on the principles of RSB.

There is the technical possibility of expanding some oil crops more intensively in palm oil, coconut, soy, sunflower, castor and jatropha, to produce biodiesel respecting the principles established in the LPDB and considering that the marginal costs of biodiesel production are increased as the aptitude of the land is reduced. The increase in second-generation biodiesel production can be obtained from the abundant available ligno-cell resources, such as agricultural residues or agroindustrial and forestry co-products, when the conversion technologies have reached commercial maturity and become competitive. The conversion infrastructure of crude oils to biodiesel, this is limited for the moment to a small number of demonstration plants of very small scale.

The ventures for companies producing biodiesel in Mexico for the years 2008-2009 that were their beginnings in the Mexican market, there was a confusion and instability, generated uncertainty that caused that the enterprises in this sector were not encouraged, so It was necessary an intervention of formal institutions, and the state who can generate new regulatory frameworks in any sector.

After the review of events on how to change the landscape of institutions in Mexico to be a first-time trial and error, it has been possible to consolidate and strengthen entrepreneurship initiatives in biofuel companies, especially biodiesel has been favored with a growth of twice what existed. So it can be noted that if there is a relationship between the formalization or solids of the institutions with the initiatives and start-up of enterprises, which supports the theoretical basis and study that was carried out by Fuentelzas, who show how institutions are an important factor for successful ventures in their areas.

FUTURE RESEARCH DIRECTION

The future research on entrepreneurship of biodiesel companies in Mexico from a perspective based on the institutions should consider the impact of the institutions in the design and implementation of a program oriented to the promotion of the most economic forms of biodiesel is considered the most reasonable short-term option term. This institutional impact must consider the different combinations of institutional incentives such as economic, fiscal, subsidies, etc. that allow biodiesel to compete in a range of costs similar to that of fossil diesel at current oil prices.

Analyze and evaluate the possibility of introducing a larger-scale program after a detailed analysis of the impact on the Mexican economy as a whole, including the direct cost of support, the positive impacts of the investment, indirect tax revenues, trade balance, generation of employment, among others.

Another line of relevant research is the analysis, design and implementation of structural programs to strengthen institutional capacities that allow the acquisition of institutional and technical experience for the development of the biofuels and biodiesel market.

Analyze institutional capacities to generate additional income in the agricultural sector and investment in production infrastructure in the industrial sector from investments in the generation of biofuels and biodiesel, with the consequent generation of employment and associated tax revenues.

Analyze the impacts of formal institutions on national and local economies by encouraging the choice of biodiesel against diesel of fossil origin and consider that in addition to the environmental benefits of reducing greenhouse gas emissions, there are other positive economic impacts derived of the introduction of biodiesel.

Evaluate the cost-benefits of adopting an institutional program that stimulates the production, distribution and consumption of technically viable biofuels and biodiesel based on the potential availability of available resources.

Estimate the institutional capacity for the production of biodiesel with national resources with reductions in biodiesel imports, in such a way that they contribute to improve the country's trade balance and to mitigate the negative effect of high oil prices.

CONCLUSION AND RECOMMENDATIONS

There is an incipient industry in biodiesel production in Mexico based on raw materials based on used cooking oils, which are accompanied by processes of processing, collection, distribution and marketing, mainly carried out in institutions that operate informally. However, because the potential for the expansion of biodiesel production is broad based on other types of inputs with increasing production trends, such as fats of animal origin and vegetable and oil-based oils such as soybean, palm and coconut, which implies the establishment of more formal market institutions to carry out the distribution and commercialization of the product. The establishment of market mechanisms in formal institutions will allow reaching higher recovery rates with lower costs, but above all, it will allow to cover in greater percentage the current demand for biodesel.

Because of this one of the concerns that must be had if the economy is to be more competitive compared to countries with a faster growth rate than the Mexican, it is necessary to be clear that formal institutions is a way in which entrepreneurs achieve organizations that exploit available markets and do so in a sustainable manner.

In addition, all those who seek to generate a project in the renewable energy sector as well as any other type of industry need to look for countries with better formal institutions, or in case of entering emerging markets that usually do not have a solid structure in its laws, regulations and regulatory agencies, to anticipate the failures that can be caused and the costs that this would cause to enter the market with proactive strategies and not wait to be in unsustainable situations.

REFERENCES

Agarwal, A. K. (2007). Biofuels (Alcohols And Biodiesel) Applications As Fuels For Internal Combustión Engines. *Progress in Energy and Combustion Science*, *33*(3), 233–271. doi:10.1016/j.pecs.2006.08.003

Aldrich, H., & Auster, E. R. (1986). Even Dwarfs Started Small: Liabilities Of Age And Size And Their Strategic Implications. *Research in Organizational Behavior*, *8*, 165–198.

Baumol, W. J. (1990). Entrepreneurship: Productive, Unproductive, And Destructive. *Journal of Political Economy*, *98*(5), 893–921. doi:10.1086/261712

Basu, K. (2013). *Más allá de la mano invisible*. Fondo de Cultura Económica.

Bazua, F. Y., & Valenti, G. (1993). Hacia Un Enfoque Amplio De Política Pública. *Revista de Administração Pública*, 84.

Comisión Reguladora De Energía. (2015). *Desarrollo Del Marco Regulatorio Mexicano En Materia Energética*. Author.

Contreras Comeche, R., & González García, N. (2010). La Medición Del Valor Social Y El Impacto De Los Emprendedores Sociales. In Emprendimiento, Economía Social Y Empleo. IUDESCOOP, Instituto Universitario De Economía Social Y Cooperativa Dela Universidad De Valencia.

Constitución Política De Los Estados Unidos Mexicanos, Diario Oficial De La Federación. (2017). *Constitución Política De Los Estados Unidos Mexicanos*. Author.

Foro Económico Mundial. (2014). *The Bold-Ones High-Impact Entrepreneurs Who Transform Industries*. Foro Económico Mundial.

Friar, J. H., & Meyer, M. H. (2003). Entrepreneurship and start-ups in the Boston region: Factors differ- entiating high-growth ventures from micro-ven- tures. *Small Business Economics*, *21*(2), 145–152. doi:10.1023/A:1025045828202

Fuentelsaz, L., González, C., & Maícas, J. P. (2015). (Forthcoming). ¿Ayudan Las Instituciones A Entender El Emprendimiento*? Economía Industrial.*

Fuentelsaz, L., & Gonzáles, C. (2015). El Fracaso Emprendedor A Través De Las Instituciones Y La Calidad Del Emprendimiento. *Universidad Bussiness Review.*

Hitt, M. A., Ireland, R. D., Sirmon, D. G., & Trahms, C. A. (2011). Strategic Entrepreneurship: Creating Value For Individuals, Organizations, And Society. *The Academy of Management Perspectives*, *25*(2), 57–75.

Hofstede, G. (2011). *Culture's consequences* (1st ed.). Thousand Oaks, CA: Sage Publ.

Ley De Promoción Y Desarrollo De Los Bionergéticos. (2008). *Ley De Promoción Y Desarrollo De Los Bionergéticos*. Publicado En El Diario Oficial De La Federación El Primero De Febrero De.

Liñán, F., Fernández, J., & Romero, I. (2013). Necessity And Opportunity Entrepreneurship: The Mediating Effect Of Culture. *Revista de Economía Mundial*, *33*, 21–47.

Mahoney, J. (2012). Property rights theory. In Economic Fundations of Strategy (pp. 158–199). Champaign, IL: Academic Press.

Marulanda, F., & Morales, S. (2016). Entorno Y Motivaciones Para Emprender. *Revista Escuela De Administración De Negocios,* 8.

Medina, I., Chavez, N. Y., & Jauregui, J. (2012). Biodiesel, Un Combustible Renovable. *Investigación Y Ciencia, 20*, 62-70.

Molina, C. (2012). Estudio de la composición y estabilidad de biodiesel obtenido a partir de aceites vegetales limpios y procedentes de aceites de fritura. Servicio de publicaciones de la Universidad de la Laguna.

North, D. (1990). *Institutions, Institutional Change and Economic Performance*. Cambridge, UK: Cambridge University Press. doi:10.1017/CBO9780511808678

OECD. (2017). *OECD Stats FAO Agricultura*. Retrieved from Http://Stats.Oecd.Org/Viewhtml.Aspx?Queryid=58648&Vh=0000&Vf=0&L&Il=&Lang=En#

Peng, M. (2012). *Enfatizando Las Instituciones, La Cultura Y La Ética*. Gestión Estratégica. Editoral Cengage Learning.

Reynolds, P., Bygrave, W. D., Autio, E., Cox, L. W., & Hay, M. (2003). *Global Entrepreneurship Monitor 2002 executive report*. Wellesley, MA: Babson College.

Rembio. (2009). *Expert Opinion Based On Delivered Questionnaire*. Author.

Riegelhaupt, E., Odenthal, J., & Janeiro, L. (2016). *Diagnóstico de la situación actual del biodiésel en México y escenarios para su aprovechamiento*. Informe Final CONFIDENCIAL red Mexicana de Bioenergía. Ecofys 2016 por orden de: Banco Interamericano de Desarrollo para SENER

Rutz, D., Thebaud, A., Janssen, R., Segura, S. A., Riegelhaupt, E., & Ballesteros, M. ... Bravo G. (2009). *Biofuel Policies And Legislation In Latin America*. WIP Renewable Energies; Report Of The EU Project Biotop (FP7); Contract No.: 213320.

SAGARPA. (2017). *Impulsa SAGARPA Producción De Biocombustibles En México*. Author.

SAGARPA. (2016). *Direccion General de Fibras Naturales y Biocombustibles. BIODIÉSEL 06052016. Presentación realizada en el Taller de Biodiésel*. Paquete Tecnológico de Higuerilla y Paquete Tecnológico de Jatropha.

SAGARPA. (2010). *Monografía de cultivos: palma de Aceite. Subsecretaría de Fomento a los Agronegocios*. Author.

Salgado, C. E. (2003). *Teoría de costos de transacción: una breve reseña. Cuadernos de Administración*. Retrieved from http://www.redalyc.org/articulo.oa?id=20502604

Sánchez, L. A., & Pérez, E. (2015). Las Entidades De Economía Social Como Protagonistas De Un Nuevo Modelo De Emprendimiento Y Medidas Legales De Apoyo Al Emprendimiento, CIRIEC-España. *C.I.R.I.E.C. España*, *84*, 35–62.

Schumpeter, J. A. (1934). The Thory Of Economic Development: An Inquiry Into Profits, Capital, Credit, Interest And The Business Cycle. Harvard Economic Studies, 46.

Stam, E., Suddle, K., Hessels, J., & Van Stel, A. J. (2009). High-Growth Entrepreneurs, Public Policies, And Economic Growth. Public Policies For Fostering Entrepreneurship: A European Perspective. International Studies In Entrepreneurship, 22, 91-110.

Tapia, P. (2017). *Biodiésel, a la espera de brillar en México*. Retrieved from http://www.milenio.com/negocios/biodiesel-a-la-espera-de-brillar-en-mexico

Urbano, D. Y., & Díaz, J. C. (2009). Creación De Empresas E Instituciones: Un Modelo Teórico. In Creación De Empresas. Aproximación Al Estado Del Arte. Lisboa: Juruá.

Urteaga, E. (2013). *La teoría del capital social de Robert Putnam: Originalidad y carencias. Reflexión Política*. Retrieved from http://www.redalyc.org/articulo.oa?id=11028415005

Vargas, J., Guerra, E., Bojórquez, A. Y., & Bojorquez, F. (2014). Gestión Estratégica De Organizaciones. Elaleph, 134.

KEY TERMS AND DEFINITIONS

Biocombustibles: It is a mixture of organic substances that is used as fuel in internal combustion engines. Derived from biomass, organic matter originated in a biological process, spontaneous, or provoked, usable as a source of energy.

Biodiesel: Biodiesel is a liquid biofuel that is obtained from lipids of natural origin, such is the case of vegetable oils or animal fats that have had or not previous use, subjected to the transesterification process.

Company: A company is an enterprise or business that has been organized as a business and commercial society including capital and labor and whose main purpose is to obtain an economic benefit.

Entrepreneurship: It is the ability of people to create new businesses. It is the person who knows how to discover, identify a specific business opportunity and then will arrange or get the necessary resources to start it and then take it to fruition.

Formal Institutions: They are the institutions related to constitutions, codes, laws, contracts, and other legal elements.

Informal Institutions: Are extensions, interpretations and modifications of formal rules, rules of behavior, agreements, codes of conduct or conventions and all those aspects that are related to culture.

Institution: Institutions are constraints that arise from human inventiveness to limit political, economic and social interactions. They include informal restrictions, such as sanctions, taboos, customs, traditions, and "codes of conduct," as well as formal rules (constitutions, laws, property rights).

México: A country located in North America.

Section 3

Specific Cases on Sustainable City Development

Bio economy has a high importance for generating environmental, ecological, and socio-economic impacts, especially in populated urban areas and peripheries. This last section presents four chapters related to the topic of sustainable city planning and development. Each chapter explores and analyzes a specific city: Warsaw, Guadalajara, Chinese Villages in Guanzhong, and Tokyo.

Chapter 6
Ecological Centre of Warsaw as a Development Path

Krystyna Gruszecka
Warsaw University of Technology, Poland

ABSTRACT

The purpose of this chapter is to present a new development path towards greening the city center of Warsaw in the context of the increasing density of building development in recent years. After the process of urban sprawl, there has been a tendency to concentrate buildings, especially in the city center. Currently, a new idea and the need to improve the climate in the city is being born. The aim is to convert dispersed green areas into a continuous network of corridors and green spaces, comprising the city center on both sides of the Vistula River. The future structure of green areas in the central part of Warsaw will be built as an element of bioeconomy. According to this new pattern of urban greenery, larger green enclaves will be connected by corridors created out of necessity along densely built-up streets. Such elements as green walls, green roofs with decorative greenery and food crops, pocket greenery, as well as urban farms (e.g., algae energy generation) will complement buildings, foster healthy environment, and create the opportunity to enjoy pastimes.

DOI: 10.4018/978-1-5225-7958-8.ch006

INTRODUCTION

20 years ago, in the 1990s, development with regards to green areas was treated as a natural asset of cities of all sizes in Poland[1]. In the 20th century, Warsaw had a radial system of green wedges converging in the centre.[2] The Study of Conditions and Directions of Spatial Development of the Capital city of Warsaw, adopted in 2006 in the new reality of the democratic system, apparently extensively allowed for construction projects in these green wedges, with the exception only the Vistula River valley together with the adjacent open green areas. The Vistula River valley played a significant role in ventilating urban built-up zone.

Modernist principles of building urban structures in the form of areas of residence, work and rest, the latter which were most often dominated by greenery, provided the standard of urban activities. During communist rule until 1989, there were limitations in terms of the possibility to register in the capital new residents from outside Warsaw. Freedom of movement that appeared in the 90s immediately caused urban sprawl - migration of Warsaw residents to suburban areas in search of areas of residence with better health conditions, preferably in proximity to the forests surrounding the city. On the other hand, new people started settling in Warsaw, especially its downtown, mainly in search of permanent employment or in order to rent a flat.

Methodology

The research methods adopted in this chapter are aimed at gathering and analyzing sources, including planning documents, available publications, individual designs directly in the field, as well as results of competitions and scientific theories, which have been successfully applied in other Western European countries. The current trends in urban greenery planning, several revitalization projects, both completed and awaiting their implementation, have also been described.

The following *course and sequence of description of the issues* have been established:

- The distribution of parks within the present center of Warsaw: contemporary, historical and already constituting national heritage, squares, protected areas, such as Natura 2000 - areas belonging to the network of protected natural areas in the territory of the European Union, in order to preserve specific habitats and valuable species of wildlife. Warsaw is the only city in this part of Europe, in which Natura 2000 areas are located in in the city center;

- Principles and examples of incorporating green enclaves into a network system comprising the city center and connected with other districts - methods and specific actions (street greenery / rows of trees, a footbridge across the Vistula River, pocket greenery, green roofs and walls as well as green energy-generating urban farms, also in the form of air-purifying drips);
- Creation of a greenery system in the city center, combined with other selected service functions to provide new conditions for the development of the city - spaces eagerly visited by residents and tourists;
- The starting point for the above considerations will be the analysis of the existing functions supporting the greenery system (summer Chopin concerts in the Ujazdowski Park, presentations of the Warsaw Water Filters during the Night of Museums as a great monument of urban engineering culture from the late nineteenth century, National Library in the Mokotow Field - the historical airport for gliders, sports fields and facilities, meeting places for the local community, integration centers, dining facilities in these areas and other income-generating, and, at the same time, attractive offerings, which also contribute to the creation of user-friendly space.

The final conclusions concern specific changes in the development of the green areas in the center of Warsaw, aimed not only at improving health conditions, but also at achieving economic successes by selecting the appropriate accompanying program, but also leading to the creation of the attractive spatial form of the new development attracting new users of the area. All efforts for and contributions to bioeconomy of the future will be highlighted.

Why Is This Chapter Devoted to Warsaw: The Capital of Poland?

Poland is the country that covers the largest area (312 000 km^2) in Eastern Europe. The second largest country of this Eastern block is Romania (however, with an area of 238 000 km^2).[3]

The population in Poland reaches nearly 38 million, while in case of Romania, for example, the figure is 22 million. In other countries of Eastern Europe, the population mostly amounts to below 10 million inhabitants. With such statistics and taking into account the location of Warsaw on the line between Berlin and Moscow (always being used the way for Polish troops marching westwards and German troops eastwards in cases of their mutual aggression) it is easier to understand that since the city suffered up to 87% damage at the end of II World War, over the last century

Warsaw was rebuilt in its entirety and expanded. Moreover, after the liberation from the communist regime in 1989, the development of this city became violent, partly uncontrolled. Today, the number of permanent residents of Warsaw amounts to nearly 1.8 million, and, according to professional forecasts for 2040, prepared, among other data, on the basis of new and planned new job forecasts, mainly the city center and the outskirts of Warsaw are to grow by another 2 million people.

The activity increase in the center of the capital city is taking place in the process of function, as a result of development of all available post-industrial, post-construction and brownfield areas, as well as poorly built areas, areas with low intensity (greenfields), sometimes, unfortunately, green areas in the form of existing squares.

Warsaw, among 22 cities in other countries, is already perceived as the world's metropolis, it displays similar characteristics as cities of the alpha type, but exerts a smaller range of influence. The remaining Polish cities are essentially different in terms of the number of inhabitants. Krakow, the city appraised the most among them, is a magnificent historic city, historically the first capital of Poland. It is seen as a metropolis of the gamma type. Two other cities, Wrocław and Poznań are waiting to reach for similar honourable evaluation in the ranking of the Global Cities Index.

Due to the size, constantly increasing population and the increasing density of buildings, as well as owing to the prestige of the capital of the largest country in former Eastern Europe, today a member of the European Union, Warsaw stands out among other cities. It is eagerly visited by Polish and foreign tourists, which opens field of activities towards the bioeconomy, especially in the area of the downtown. The issues concerning climate protection in this zone are much more complicated here than in case of smaller urban centers. That is why this chapter was devoted to creation of ecosystem network, at least on the scale of the city center, that would create a continuous system of riverside greenery, parks and other contemporary solutions improving the quality of life in this area.

At the Turn of the Century

In the 1990s and at the beginning of the 21st century, there was a tendency to scatter buildings away from the centre to other areas of the agglomeration in search of more attractive, more ecological living conditions. In recent years, buildings with important functions for shaping the Warsaw metropolitan character have begun to return to the central parts of the city, while the number of residents there has been slightly decreasing. New concentrations of office buildings, services and investments important for the emergence of Warsaw in the global world cause the reduction of greenery as part of new housing and services developments.

As a results, the climate of the city is deteriorating, air pollution is increasing, and smog harmful to health occurs constantly during the heating seasons. The overlap of these problems with the publicly unexplained and probably unjustified mass felling of trees in the city in 2017, also in the squares in front of important public facilities (such as the Śródmieście/Downtown District Office), resulted in a positive social and institutional response, where all stakeholders teamed to improve this unfavorable situation, which already constitutes a threat to the health of Warsaw residents and permanent users of the city center - the area most threatened by the effects of environmental deterioration.

Creating of Urbanity Node With Bio-Economic Characteristics

This chapter aims to present the main objectives of new activities planned by the local government authorities, also within the framework of the new administrative unit - the Urban Greenery Board of Warsaw (Zarząd Zieleni m.st. Warszawy). The Board is supported by Deputy Mayor of Warsaw and encouraged to apply the latest pro-ecological and innovative solutions used in the fastest developing cities of the world.

The numerous activities presented in this text are supposed to constitute a greening strategy for improving the health of residents. The investments planned for implementation until 2020 set the directions for improving the living conditions in Warsaw - a rapidly growing city. There have been many efforts to ensure that these activities are not only ecological, but also economically justified. Above all, these are the activities for creating traffic generators - places which attract the largest number of space users.

These should be attractive spaces that encourage people to spend time in them and use all facilities, innovations, educational elements, opportunities for recreation and practicing various forms of sport. A good environment for existence, work, rest, play, shopping and other similar activities brings about the synergy effect and also attracts other functions such as catering, the need to buy clothes suitable for sports, cultural functions (exhibitions, museums, concerts, specialist and general courses).

Bioeconomy is excellently supported by these additional forms of activity. A so-called *urbanity node with bio-economic characteristics* is created. This type of activities in Warsaw will be described in this chapter as a methodical principle for implementing financially profitable investments and improving livability of spaces.

URBAN GREENERY DEVELOPMENT

Recent analyses of the capital city carried out by scientists and officials, as well as surveys among residents and other social groups using urban space show that knowledge about pro-ecological development activities of urbanized or developing areas is widely available and partially known. The problem, however, is the management of activities which could be used for harmonious, sustainable development.

HISTORICAL PARKS IN WARSAW: POTENTIAL FOR CHANGE

Most parks in the center of Warsaw are historical gardens, which are now listed in the register of monuments. In the Park of Agrykola, revitalized in 2008, a vast panorama with the characteristic axis of the Piaseczyński Canal extends from the upper terrace under the Ujazdowski Castle to the Vistula River.

There are also special Chopin benches in the Saxon Garden (Polish: Ogród Saski), the Royal Baths Park (Polish: Łazienki Królewskie) and in 13 other places in Warsaw, which are a kind of contemporary monuments of this famous composer and his work.

Sitting down on such a bench, you can listen to Chopin music, find out why he created his pieces in a particular place of the world and find the location of other such benches on a special map. This innovative solution attracts fans of romantic music and delights users with wonderful atmosphere. The location of the benches in the ecologically active city area creates additional attractors in the form of beautiful green landscapes, which, of course, translates into greater flow of people using these areas and other economically justified facilities located there (monuments, restaurants, souvenirs).

Next, the Ujazdowski Park is considered a neat, old, properly looked after place. It is possible to relax there on the grass or, for example, to listen to a concert in a clearing near the monument of Fryderyk Chopin, while sitting on the green lawn.

The monument was first built in August 1901. Poland was then *partitioned,* with Warsaw under the Russian occupation[4].

After the war, the bronze statue of Chopin sitting under a willow tree was rebuilt. It is there until today, and in summer, crowds of people sit on the grass listening to piano concerts held *in situ* at the monument under the "weeping" willow with bent branches - which is a highly symbolic tree in Poland. Live concerts of romantic classical music in the open air are thrilling, just like the walks down the park paths from the escarpment to the lower level of the Royal Baths Park with picturesque bars and cafes, the historic Palace on the Water in the background, or the orangery housing tropical plants. Classical arrangement of the park is enriched with various water forms, unique plant species, and also with historic architecture: pavilions, the Royal Summer Palace (a monument and sightseeing attraction), Amphitheater on the Water (venue of theatrical plays in the open air) and others, such as the Myślewicki Palace, Old Orangery, White House, Hermitage, Egyptian Temple, Belvedere. The audience of Chopin concerts are economic consumers of the listed attractions, starting with the Chopin concerts themselves - and because there is no admission fee, the concerts are available for everyone willing to listen and relax, and continuing their use of the park in its interesting and varied infrastructure.

Bearing in mind that contemporary people are more in contact with machines than with other human beings - several hotspots have been created throughout the park for those in need of Internet access.

The Palace on the Water, which was the summer residence of King Stanisław II Augustus in the second half of the 18th century, is considered one of the most beautiful palaces of this period in Europe. This place is used as a venue for various cultural events, lectures, films, interviews and discussions. In the summer, Warsaw participates in the so-called the Night of Museums project, that is, the night when all interesting facilities are open to visitors free of charge. Crowds of art and architecture fans go from one historical location to the next, admire the presented monuments, and if they are interested, they come back there for various events on other summer days. The Night of Museums is the initiation of many people's interest in culture, art and monuments.

Another gem in Warsaw is the Krasinski Garden in the Muranów district with a vast grassy foreground in front of the Krasinski Palace and lush green surroundings. The Saxon Garden adjacent to the Józef Piłsudski Square is characterized by the axial composition starting at the Tomb of the Unknown Soldier and continuing to the west. Ahistorical identification / perception of these parks raises the problem of the manner in which they are planned and used.

For example, a visit to the historical gardens of Paris evokes a communicative image of these green compositions, taking the visitor to the world of the past - as if taken from centuries ago, impressive with their multitude and variety of forms, usually with varied trimmed greenery. Warsaw gardens grow luxuriantly, nobody plans or shapes them in reference to past designs - they simply exist, and often the landscapes of historical gardens become similar to those of today. Considering that modern parks established in the middle of the 20th century also function in the city center of Warsaw, the use of knowledge about the historical shape of green compositions of these centuries-old gardens would be advisable - for their popularization, educational purposes and greater social pleasure in such space.

Undoubtedly, these remarks can be treated as suggestions for further revalorisation activities, also for more effective economic use of these historic sites, in which more or less high-end places could also be created, with more expensive services complementing park landscapes and numerous cheaper, well-designed ones. The rationale for the fact that the historical design of monumental park complexes was not used everywhere in Warsaw is undoubtedly the total destruction of the city center of the Polish capital during World War II. The Germans destroyed 85% of the buildings in Warsaw, cut out the forests around the city, and destroyed the city parks.

The problem of combining "bio" activities with economics is not always fully understood and appreciated by professionals, sometimes by domestic green activists, but also by regular residents. Such a situation occurred in 2013 when a competition was organized for the 21st-century garden design in the Royal Baths Park in Warsaw. This project was called the space of synthesis, consent and building relationships. The philosophy of the planned garden was closely related to the contemporary world trends in this century.

This garden reflected the characteristic relationship of man with nature. At the same time, it was an image of contemporary aspirations and hopes in the 21st century. The intentional message of this work was: „protection of nature - protection of life". The aim of this concept is to increase ecological awareness in the name of man's respect for nature and quality of life improvement. We are dealing here with the interpenetration of architecture and greenery. It is also important that the object itself was to be energetically self-sufficient, both generating and harvesting energy, just as it happens in nature; it was also to act as a sewage treatment plant. This was the concept and objectives to transform this idea into an economic result which could later be used in other cases.

Taking into account the shortages of exhibition space in the Polish capital, the Exhibition Pavilion Complex was proposed as part of this architectural development, so that Warsaw could more fully integrate into the current trends of international exchange of museum and artistic exhibitions.

These pavilions were to be covered with earth and greenery, with only the entrance part in the architectural form. Such a spectacular spatial solution was to attract visitors to this unique place (Figure 1).

Figure 1. The design of the exhibition pavilion for an investment in the Royal Baths Park called "The Garden of the 21st Century" (https://www.toposmagazine.com/garden-21st-century/#!/foto-post-130-1)

As a result, the Royal Baths Museum - already a natural complex with many interesting buildings - would become an important center of art, European education and international cooperation. The project belonged to the landscape trend characteristic of Polish regionalism.

The project used green walls, which constituted an extension of the biocenotic garden and created an image of non-material architecture. It was an example of High Art, and it also had educational functions of promoting new technologies and creating exhibition space for modern solutions in connection with nature.

In terms of the development structure, flexible spatial arrangements of pavilions were used, allowing any connection with the foyer or division of the space around the green ring on the axis of the New Orangery from the north side, and the garden amphitheater on the opposite - south side.

After several years of protests, it was proposed in 2017 for these spectacular exhibition space with an area of 1800 sq m, to create space on "minus 1" level with a natural carpet in the form of a flower meadow on the roof to satisfy the critics of this original project. When the museum staff, who were the initiators of the idea, were exchanged, the conference and exhibition center concepts remained in the planning stages, available also on the Internet.

This is an example of how the cultural space of a historic park could be revived, and yet those dreams about the area for artists and other exhibition purposes were abolished by surrendering to social movements which inhibited these initiatives from proceeding. The memory of the project remained; nowadays, there is a growing consensus among decision-makers, who are forced to listen to the voice of the public, and representatives of protesters, who express the needs of the local community. All stakeholders often manage to reach compromise.

There is a chance to return in the near future to this extremely innovative project, opening the prospect of harvesting the energy we need in a natural way, and, at the same time, we should perceive it as giving social, educational and simply economic benefits.

When considering how to preserve the existing condition and what actions to take to embellish and enliven these historical green areas, it is worth taking into account the historical layouts of these gardens. In many cases, these were not only recreation areas developed as greenery with pedestrian alleys and benches for resting. They served various additional purposed or had the equipment which offered added value - something to attract people willing to be active.

It is worth mentioning the Mineral Water Pump Room in the Saxon Garden, where ornamental tables and chairs with rich patterns made of bent metal rods were placed in summer, and ladies in long dresses were walking with their high-hatted companions, sipping wholesome mineral water through a straw or from a cup (Figure 2, 3).

Also, there were decorated hand-pumped wells in the metal fencing, where visitors could pour a cup of fresh healthy water themselves. Apart from these attractions, the Summer Theater was built in a different part of the Garden - a wooden building symbolizing good entertainment, music, as well as social and cultural events.

These facilities are no longer present in the Saxon Garden (Figure 4) they disappeared just like the magnificent Orthodox Church - a symbol of foreign domination, destroyed in the 1920s after the First World War, when Poland was already an independent state.

Currently, there is no such space planning which automatically stimulates the flow of people of various tastes through the park. The result is drowsy emptiness of this green area. According to the results of surveys carried out also on Facebook, people who incidentally rest there, do not usually stay there longer than 45 minutes.

Figure 2. The Saxon Garden in the center of Warsaw - the establishment of the Saxon Axis from the end of the 18th century; image from 1918. Important non-existent facilities are marked: A - Saxon Palace from the end of the 18th century, B - the Brühl Palace, C - Orthodox Church.

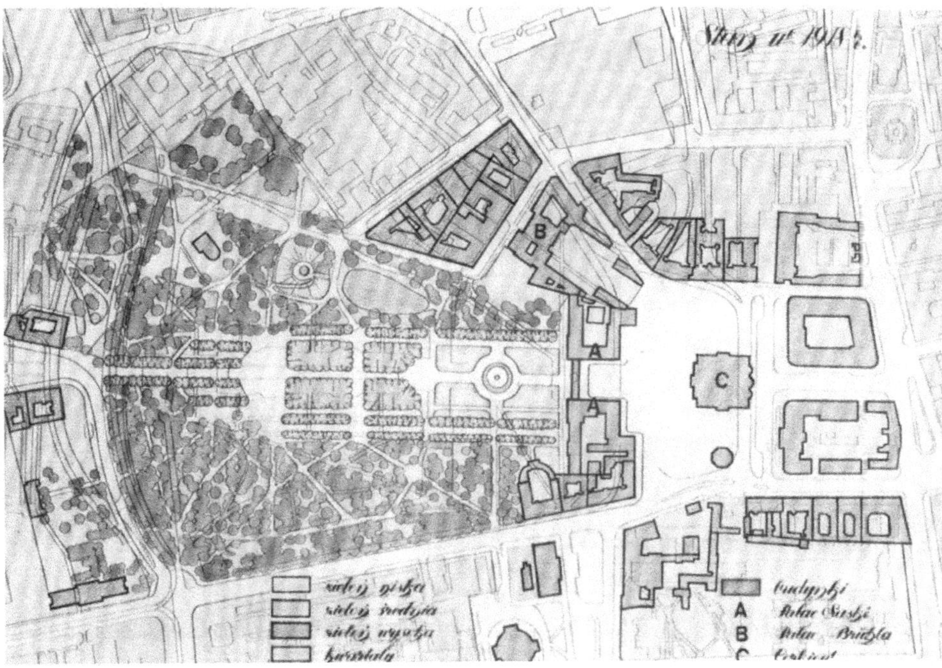

In recent years, there has been a great discussion in the Warsaw media whether to rebuild or not rebuild these palaces. The author of this text created a concept of urban reconstruction of the historical layout of this area in the 1990s, cooperating with a company of investors from Switzerland. As the palaces of that time belonged originally to the electors from Saxony, then to the Russians, they were sometimes associated with foreign domination, which hindered the decision-making process whether to rebuild them. However, this place positive marked its existence in the history of the Polish nation. At the beginning of the 19th century, at the Congress of Vienna, the Saxon king Frederic Augustus I, who owned the Saxon Palace, was forced to resign from the title of the Warsaw prince. As a result, the Saxon Palace was sold to the Kingdom of Poland. In some rooms, the Warsaw Lyceum was located. In the years 1810-1817, young Fryderyk Chopin lived with his parents and sister in the right wing of the palace.

Figure 3. The Institute of Mineral Water in the Saxon Garden in the form of a publicly available pavilion (http://www.warszawska.info/parki/ogrod_saski-historia_dawna.html)

Figure 4. A perspective view of the main body of the Palace, planned for reconstruction from the cour d'honeur side
(hand-made sketch by author, 2013)

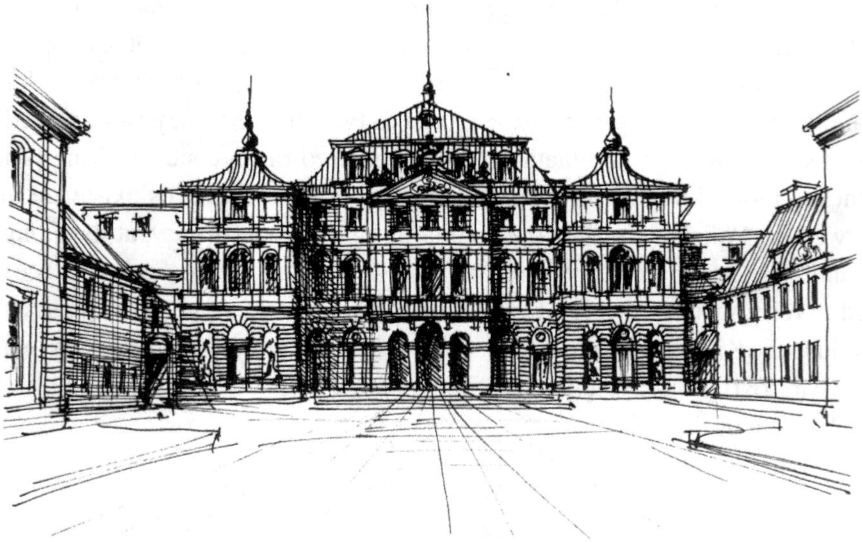

Another exceptional event took place in 1979. Ten years before the political changes in Poland, Pope Karol Wojtyla addressed the nation speaking in the Saxon Square - millions of people gathered there, also in the Saxon Garden. John Paul II encouraged hope and gave strength to continue the struggle to regain freedom and join the democratic world. His words were invaluable for Poles (Figure 5).

The cited examples clearly show that this region with the unfinished Piłsudskiego Square (formerly Saxon Square) may in the future become the place for activities connected with bioeconomy. There are opportunities to strengthen initiatives which bring financial profits, such as in the cited project study - hotels, office buildings, administrative buildings, conference centers, space for international business and intellectual exchange. Facilities for these representative functions in the form of

Figure 5. The concept of adaptation of the Tomb of the Unknown Soldier in the form of three central spans of the former colonnade and the creation of an openwork curtain from the rebuilt colonnade behind the back (watercolor created by the author, 2013). The idea of the project - a combination of ecological values of the Saxon Garden (18 ha) with socially useful architecture: 1. Preservation of the modern monument and tribute to the fallen during World War II, 2. Creation of foot and viewing connections between the square (formerly Saxon, renamed Piłsudskiego) and the greenery of the Saxon Garden (the possibility of walking on the sides of the monument) 3. Reconstruction of the side wings of the historic palace for contemporary useful purposes: office, official, administrative or cultural or representation functions.

garden space, the value of which can also be raised, give it a unique character, referring to the historical layout of the garden. The potential of this place is waiting for future investment decisions.

CONTEMPORARY PARKS

Right in the center of Warsaw, on the north side of the Palace of Culture and Science, there is a small Świętokrzyski Park. It is valued because of peace, shade on hot summer days and the fact that there is such a place in the very center of Warsaw. It is the future location, from the side of the Palace of Culture and Science, of the moved furniture pavilion "Emilka" - one of those innovatively designed buildings from the 1970s, promoting the interpenetration of the outer and inner space of the commercial pavilion.

In addition, the pavilion was covered with a thin **plate** of reinforced concrete roof, freely shaped. Pavilions of this type received unbelievably positive reviews in the 1970s from artists from Western Europe, who considered them innovative.

The pavilion, reconstructed in the Świętokrzyski Park, will contain services conducive to good time spending by users of this area, thus the eco part of the park will function next to the commercial part, and both of them will provide healthy and pleasant experiences for the people using this space, contributing to their well-being.

The second contemporary green area in the center of Warsaw is the Marshal Edward Rydz Śmigły Park in the Powiśle district. The significant area of 53 ha, forming a complex of several large park layouts in the former Powiśle and the Frascati Gardens developments, is partly located on the Vistula escarpment.

The Mokotow Field (Polish: Pole Mokotowskie) located in the area of approximately 73 ha, belonging to three districts: Mokotów, Ochota and Śródmieście,[5] is a unique enclave of greenery on the southern side of the city center. Despite its fairly long and unusual history, the complex presents a stylistically coherent contemporary park landscape, designed and developed in the 1970s and 80s.

In the nineteenth century, it was the Mokotow Battle Field, comprising Russian military training grounds. In the 1880s, a track was built by the horse racing association and in 1910, the Warsaw Aviata Aviation Society located its workshops there. In 1915, when during the war the airport was taken over by the Germans, a hangar for zeppelins was established. After the war, famous aerial sports shows were held there.

After the war, in 1921, the first passenger airport in Poland was opened in the Mokotow Field, which as soon as in the 1930s was moved to Okęcie (the currently existing and functioning airport).

The complex of greenery, traditionally known as the Mokotow Field, is extremely important for promoting the idea of bioeconomy in Warsaw, because from the very beginning of its transformation from the airport to a city park in the last decades of the 20th century, the principles of bioeconomy have been implemented here. Stanisław Bolek was the author of the park layout in the 20th century. The basic formula of the project are attractive green and water forms enriched with various attractions, which appeal to the users to these recreational areas.

First of all, the development of the Mokotów Field is famous for its location along the Independence Avenue (Polish: al. Niepodległości), the National Library and the main archive of Polish literature - esteemed institutions in which users can work and read priceless books in a healthy and friendly natural environment, away from the hustle and bustle of the city. The park, consisting of two parts connected by a footbridge over the street, also has interesting solutions: walking alleys, including eg.. the Ryszard Kapuściński path, bicycle paths, but also designated areas for roller skating.

As a result, these places attract people of different age groups who seek rest. Interesting landscape, which from some perspectives closely resembles natural enclaves of the Central Park in New York, is enriched by erratic boulders, e.g. megalithic circle, lakes and fountains, as well as duck islands, the monument of the happy dog or a gallery of sandstone sculptures - the remains from artistic workshops which took place there.

In the last few years, further development of the Mokotow Field has been more often consulted with the local community. More and more ideas have been put forward for developing the park in a way that promotes liberation of users' activity (places of social integration, cultivation of social gardens).

Competitions for the concepts of revitalization projects for the transformation of the Mokotow Field are preceded by workshops for participants. During the workshops in 2016, which took place in the National Library, the following issues were analyzed by asking questions to the participants: greenery in the Field, transport, functions of the Field, presence of dogs, events and their types, miscellaneous (consultation, 2018).

According to the author of this text, it would be very useful for further revitalization of the Mokotow Field to apply the principles of bioeconomy, so that the activities were both related to nature and brought small profits, which, in turn, could become a driving force of further activities. In modern times, when interpersonal contacts have been exchanged for the contact with electronics and machines, everything that fosters and stimulates interpersonal relations will meet with positive reception.

This year (2018), a competition for the revitalization of the Mokotow Field was held. It was won by the WXCA studio from Warsaw, famous also for the original design of the Polish Army Museum at the Citadel and the design of the last section of the Vistula Boulevards. The success of revitalizing activities in the area the boulevards has been covered by the press in Western Europe.[6] The prize was awarded for reconciling social expectations and ecology. The authors focused on highlighting the importance of the main axis - the park alley.[7] This walk will open with the Air Pavilion, one of four pavilions related to elements (Figure 6).

It will present the history of the park, but also new ecological challenges, like the threat of smog. Along this main avenue, gaming tables or places for playing boules will be located, which will be interwoven with flowerbeds. Larger changes will be introduced by linking the park with the adjacent areas of the Syrenka stadium and the Chemical and Process Engineering faculty and the main campus of the Warsaw University of Technology, among others, by a footbridge over Trasa Łazienkowska, but also by including these areas in garden revitalization activities, e.g. transforming them into a permaculture garden.[8]

Figure 6. Competition design (2018) of the Air Pavilion by the team of the WXCA Warsaw Studio. The pavilion is located at the beginning of the main axis of the Mokotow Field park avenues. (http://www.wxca.pl/aktualnosci/i-nagroda-w-konkursie-realizacyjnym-na-rewitalizacje-parku-pole-mokotowskie-w-warszawie/)

In the competition, it was proposed to create a large meadow in the western part of the Mokotow Field, next to a savannah-type area with a thematic playground and, further, a nature corner. Between these elements, a statue of Józef Piłsudski - great Pole and leader from the interwar period - will be erected. In addition, as planned in the last century - there will be a designated place for a viewing balloon. The largest water reservoir in the park, currently in a concrete basin, will be changed into a more natural form accompanied with a rain garden. At the south end of the reservoir, the Water Pavilion is proposed - a two-level building, whose facade in the summer season will be made of water curtains, forming a waterfall.

In winter, it will probably be possible to skate on an artificial ice rink. There is also a planned area for urban farms - social gardens.

As can be seen from the above-described transformations of one of the largest organized green areas in Warsaw, everything possible was done so that this diversified space would become the most attractive for recreational purposes, and oriented also towards providing services for the health of residents. Such development stimulates the market value growth of the neighboring areas and will continue to do so. It is a commonly known phenomenon that residential areas around city parks reach higher prices per 1 sq m in relation to the other areas in the city center, located near service-provision or housing developments. It will be necessary to increase the intensity of housing development around the Mokotow Field to finalize this process. This will be the next stage of investment projects in the near future (Figure 7, 8).

THE POTENTIAL OF THE URBAN GREENERY NETWORK SYSTEM IN THE CENTRE OF WARSAW

One of the most daring theoreticians of urban planning in Poland, Prof. Zbigniew K. Zuziak summarizing his considerations entitled “The strategic dimension of urban planning”, stated, among others, that the search for new formulas for the assessment of space, i.e. “developing new theories that associate the city’s economy with the economics of the environment and cultural values, should aim at situations where market value does not dominate excessively over other values determining the quality of urban life.”

Such balancing of reasons or directions of development must secure the possibility of sustainable use of natural and cultural resources. While agreeing fully with this opinion, it is worth adding that the complete lack of the element of market play in these activities is also dangerous, reducing the level of care for common goods, for

Figure 7. The design of the Water Pavilion with water curtains - waterfalls (the WXCA Sudio team) - to be implemented in the near future (http://www.wxca.pl/aktualnosci/i-nagroda-w-konkursie-realizacyjnym-na-rewitalizacje-parku-pole-mokotowskie-w-warszawie/)

example public spaces, transport system or urban greenery in any form. Managing these goods is costly, with a special emphasis on their maintenance. Therefore, balancing the proportions between "bio", pro-ecological activities for improving health, climate, air cleanliness etc. and the economic consequences of these innovative processes should be monitored, responsibly checked at shorter intervals, e.g. every year, and corrected in case of adverse results. This is in the common interest of urban communities, especially those living in growing metropolitan areas.

In terms of systemic solutions in the city center of Warsaw, it is necessary to create a network of green areas. Each of the described parks is a publicly accessible space. Therefore, these spaces should be linked in order to facilitate their use for sports purposes, as well as to satisfy the needs of those seeking rest. The best example of such a layout is the sequence of parks south of the city center, beginning with the grounds of the Warsaw Water Filters, which belong to the historic King Stanislaus axis through the Mokotow Field, the Botanical Garden, the Ujazdowski and Łazienkowski parks to the areas of the Vistula River greenbelt.

Figure 8. Planned new landscapes of the Mokotow Field (http://www.wxca.pl/aktualnosci/i-nagroda-w-konkursie-realizacyjnym-na-rewitalizacje-parku-pole-mokotowskie-w-warszawie/)

City authorities took care to widen the relatively narrow strip over the Vistula River, especially that it is parallel to a transport route of inter-district importance. This expansion was achieved e.g. by creating a park on the slopes of the Citadel, the famous garden with greenery and ponds with fountains at the foot of the Royal Castle, or the square in the Powiśle boulevard between the Poniatowski bridge and a faience factory. Today, when the Urban Greenery Board of Warsaw (Zakład Zieleni m.st. Warszawy)[9] has finally started revitalizing the Vistula boulevards, intending to green their hard, stone floors, these boulevards are beginning to be a real attraction of at least holiday recreation and their accessibility becomes really important. Regular residents have difficulty getting easily and directly from the upper terrace of the Vistula escarpment to the Vistula River.

Overcoming these types of barriers is a challenge faced by designers and contractors in the coming years if the modern slogan "the city for people" is to be treated seriously, as expressing the direction in which the center of Warsaw is to be transformed. This can be done e.g. by creating green bridges over the transport route with elevators in the area of the Vistula Boulevards. Another possible solution is lowering the route, which has already been partly done. Maybe this concept should be continued, although it is technically difficult.

The fragment of the street connecting the Mokotow Field with the Ujazdowski Park should be adapted to the needs of walkers and cyclists, by crating pleasant, sensuous environment. In particular, there must be green walls and roofs, pocket greenery e.g. in the form of glass structures filling small gaps in compact development structure, facilities with cultivated farms, also with edible plants - a modern trend widely present e.g. in the cities of Great Britain. This would be the right place for unusual decorative plants along sidewalks or bicycle paths, eco-gastronomy, healthy food, drinking in a healing environment (e.g. located near life-giving moss screens or algae air purifiers). it is crucial to choose the street which is the link between larger green areas in order to attract customers to this place.

If a green outer wall is to survive the cold winter in Poland, when for several-day periods temperatures fall to minus 20 degrees Celsius, it must have a special construction, irrigation and nutrition of plants, as well as protection of roots from frost. When adapting technologies used in the countries of Western and Southern Europe, these walls are not able to survive winter conditions. The attached illustrations show both examples of implemented examples of Polish solutions and innovative ideas (Figure 9, 10, 11).

Figure 9. A green inner wall with lush vegetation in IQGARDEN technology in the H & M Polska retail outlet
(photo: Łaskarzewska M., 15/07/2018)

Figure 10. Design of a boutique hotel on Ząbkowska Street in Warsaw
(authored by Filip Orliński and Małgorzata Łaskarzewska)

Figure 11. Green gabions in IQGARDEN technology in the square next to the Zodiac in Warsaw. Square design: Bartek Gowin and Krzysztof Suita
(photo: Łaskarzewska M., 25/07/2018)

In addition, the inspirational idea was presented of oxygenating urban spaces fogged by blue and gray smog, starting in late autumn. Such devices are promoted and will have to become part of our reality.

In the city center of Warsaw, according to the latest theories on shaping the structure of central metropolitan areas, it is necessary to return to the network system - the nodes of activity and their interconnections. This system functioned in the era of the socialist economy, practically until the end of the 1980s, but with less traffic and smaller possibilities of various forms of development. There was a system of green wedges converging concentrically in the center. Currently, rapid development, including increasing density of the Warsaw city center through creation of new hubs (clusters of office buildings, concentration of services, mixed-use skyscrapers islands, important metropolitan functions of international exchange) has led to the elimination of these interconnections and building development was introduced in these green wedges.

In the network system, it is important to take into account the connections between pedestrian crossings and greenery between the upper terrace of the Vistula escarpment - the major part of the Warsaw city center - and the Powiśle district, the city area on the lower terrace - closer to the Vistula River. Figure 12 shows the principle of shaping these connections, but detailed solutions need further analysis.

The analyzed section of the city center of Warsaw should have at least a few such descents to the Vistula embankments. This is encouraged by the gradual process

Figure 12. Vision of networked green areas in the city center of Warsaw with distinction of necessary supplements of this system with additional installations (created by the author, 2018)

of revitalization of this area. On sunny weekend days, crowds of people enjoy spending time by the river. This is just the beginning of adapting the banks to the needs of residents. The revitalization program covers the entire river banks within the administrative boundaries of the city. Already, the German media are describing Warsaw and its popularity with foreign visitors, who enjoy spending time looking over the Vistula River at the panorama on the opposite side of the river to the city center, and also visit rapidly multiplying charming cafes, bars and restaurants in the adjacent Powiśle district. Therefore, the bio-economic synergy works: easy accessibility of the river and beautiful picturesque landscapes are conducive to activities which bring tangible profits to the owners of retail and service outlets.

Another problem connected with improving the structure of the city centre of Warsaw is are better links between the left and right bank of the Vistula River. It is especially important since in the Praga district, located on the right bank of the river, the greenbelt is wider and covered by the Nature 2000 network. In a competition held this year, a pedestrian bridge design was selected, connecting the Praga district with the lower part of the Warsaw city center on the left side of the river (figure 12). Undoubtedly, it will be a highly frequented place, used for recreational purposes, and an excellent viewpoint to historic Warsaw.

DEVELOPMENT OF BIOECONOMY IN WARSAW

The task of preserving and developing the quality of the most important and renowned garden and park areas is planned in the first stage of the works undertaken by the Urban Greenery Board of Warsaw. These works should also cover smaller squares located in many places of the city. According to the popular principle of "pocket greeneries", these places should be carefully developed. It is an interesting idea is to build for this purpose glass or simply transparent facilities or closed installations, which inside may have different kinds of greenery, properly and sometimes mechanically irrigated. The plants might be visible from the outside even in cold, snowy winter weather, while the proper arrangement of the interior of these facilities would provide good and safe conditions for vegetation.

A systemic approach to the issue of city greenery, taking into account the values influencing the urban layout, leads to the conclusion that a continuous system rather than a dotted arrangement of urban greenery is a better solution. City parks and squares should therefore form a network of green spaces which would allow to move from one place to the other in sequences - cycling or walking along ecologically designed passageways, squares or simply better-designed streets with ecological means of public transport, away from the traffic and pollution. This is the objective to be pursued [author's model of the system for Warsaw downtown].

Discussion

The discussion taking place among various professional groups in Warsaw concerns various trends, in relation to distress caused by the visible climate change. Creating a network of transport systems, and in recent years also creating networks of sequences and spaces covered with greenery, is currently becoming the topic for many papers and public speeches at conferences in scientific environments[10]. The system of walking and cycling routes with the accompanying greenery in Strasbourg provides an inspiring example of this. Moreover, in Strasbourg even tourist trips around the city are held in rickshaws in this type of areas. Strasbourg, however, is a much smaller city in comparison to Warsaw, as well as it is one more permanently embedded in Europe's economic systems. Warsaw is yet an emerging metropolis of Eastern Europe.

Developers from many countries, most often western European companies, sometimes from the American continent fight for places for their further investments in the downtown of Warsaw. It was not until some of these investments proved unfavorable for the living conditions and made it harder to move around the central zone of Warsaw, that the city authorities experienced an intellectual "awakening", which was actually stimulated rather by the need to protect their positions against social protests than by the desire to introduce new initiatives. The turning point was the previous year, 2017, which saw wild actions of cutting out many thousands of trees, supported by government authorities (Ministry of the Environment), as well as the Voivode of the Mazovia Province.

This situation was made possible by the amendment to the Act. Previously, the regulations determined that if it was necessary to cut down trees for new investments, such as a gas pipeline or new buildings, for example, the investor had to make sure that the number of trees that had been cut out will be planted elsewhere. In 2017, The Voivode exempted investors from this obligation. On a beautiful sunny weekend in 2017, the trees growing in the square existing there for many decades disappeared from in front of the building of the Warsaw Downtown District. Also, on the opposite side of the Vistula, a part of the greenery in the Zoo was removed. This caused a media "storm" to break out. Territorial authorities changed their position, they began to listen to the voices and opinions of residents and users. Deputy

President of the Capital City of Warsaw, Michał Olszewski, is currently a patron of many of the actions described in this chapter, he is present while all important decisions regarding the issues of shaping the Warsaw ecosystem are made. On his *Facebook* account, he awaits correspondence from every person interested in the issue. He also actively engages himself in the discussion, calmly listens to criticism and, upon checking whether it is justified or not, he draws conclusions for further positive actions from it.

The discussion on the face of the greenery system in the rapidly developing Warsaw for the last 25 years is not only a theoretical, scientific discourse. These are actual actions of decision makers in the city, which, if they fail to meet the expectations of improving the quality of life in the city, are corrected.

For just a few recent years, attempts have been made to introduce green walls and roofs to the Polish market. Such installations which, if broadly used would enrich the city in an attractive way, would improve climatic conditions, health conditions and comfort in public spaces equipped with such forms of development.

In the legislation regulating the issues of local spatial development plans, up until the last year only ground (soil) and, in a small percentage, flat surfaces of terraces adequately equipped with a plant bed were recognised as biologically active areas, that is such on which natural vegetation of plants accompanying the development can be provided. Krystyna Guranowska-Gruszecka, the author of this text, together with Małgorzata Łaskarzewska, the co-owner of the patent for the Polish Green Wall, actively participated in the application process, carried out in the Ministry, in reference to the amendments introduced to the basic spatial planning law last year. By means of doing so, they secured the complements that enable architects to arrange greenery also on vertical architectural elements. Arch. Łaskarzewska states that thanks to this, it has been made possible to apply *"research and implementation research in the Green Technologies Consortium as part of the NCBR Biostrateg grant project" to the urban space.* The development of vertical solutions related to the use of a biodiverse system for the decoration of walls and roofs of buildings built in cities is the subject of the following research. The owners of the Polish Green Wall patent, established under the INNOTECH grant, the IN-TECH path as a result of the research, proposed to extend its application with the possibility of forming a green wall with water retention. Since January 1, 2018, appropriate regulations have been introduced that allow for such implementation. The first investments of this type have already been created and are being tested in the very center of the city. Thus, the idea and technology appeared to build such walls in highly urbanized tissue where it is impossible to build horizontal areas of greenery. To quote Małgorzata

Łaskarzewska: "This complex of multilateral gabion walls was created in the center of Warsaw by the Zodiac, the building of the Warsaw Association of Polish Architects. In 2019, it is planned to complete another implementation of green wall at the famous Rotunda with a green-walled/with a green patio, the top investment for Warsaw Downtown."

Bartosz Dankiewicz summarizes the advantages of this installation, which *"improves the purity of air, provides access to clean water reservoirs in the city center, and, based on rainwater, it uses municipal water supply to a small extent, which clearly enhances the city's topoclimate."*

Since such installations are implemented in densely built-up areas of inner city of metropolis, the aesthetic values of these regions increase, which undoubtedly contributes to the recognition of these places as unique. Thanks to this, these parts of the city will be visited by residents and tourists more often.

Having such a proven technology coverage in Polish difficult climatic conditions of cold winters, K. Guranowska-Gruszecka proposed an original model of a network of links to the green areas described here, including parks, fragments of street sequences equipped with a system of green walls and roofs. The idea was to create such an ecosystem for pedestrians and cyclists that, as one can imagine, will successively attract economic functions, namely gastronomy, art exhibition, culture, sport, entertainment and suchlike. The appearance of these functions will prove decisive for the profitability of these investments planned as part of the city's bioeconomy system.

Positive Changes

The objective of the Urban Greenery Board of Warsaw (Zakład Zieleni m.st. Warszawy) is to green 74 city streets, starting with those of supra-local functions. In November 2017 10 procurement procedures had already been finalized and contracts awarded for developing the concept of these useful activities.

In July 2018 the first streets and squares have already been improved with lush greenery. Considering that this new administrative unit, the Urban Greenery Board of Warsaw, was established in order to implement the program for the years 2016-2020, there is a chance that subsequent streets will also undergo a "green" change. Świętokrzyska Street in the city center is a positive example of this implementation. After a new metro line was built under the street (2017), the width of the carriageway was reduced, bicycle paths were introduced and impermeable floor slabs were laid on very wide sidewalks. The created space that was too vast and empty, and, in effect, seemed inhuman. The revitalization created surprisingly friendly landscape and atmosphere for walks (Figure 13). Even retail outlets located on the ground floors of the buildings fill up with interesting offers.

City streets vary in terms of their role and conditions: their role communication for vehicles an pedestrians, their width differs, they are equipped with technical infrastructure in various ways, sometimes underground or railway tunnels run underneath the streets, which significantly limits the possibility of planting trees, which need space underground for their root.

This will make it necessary to create standards for the improvement of streets with greenery under different conditions. It is assumed that these will be such trees as oaks, plane trees or lindens. It is also allowed to plant shrubs alone, without trees or alternately with trees - in streets where, for example, underground infrastructure

Figure 13. Model of green network for the Warsaw downtown
(K. Guranowska-Gruszecka 2018

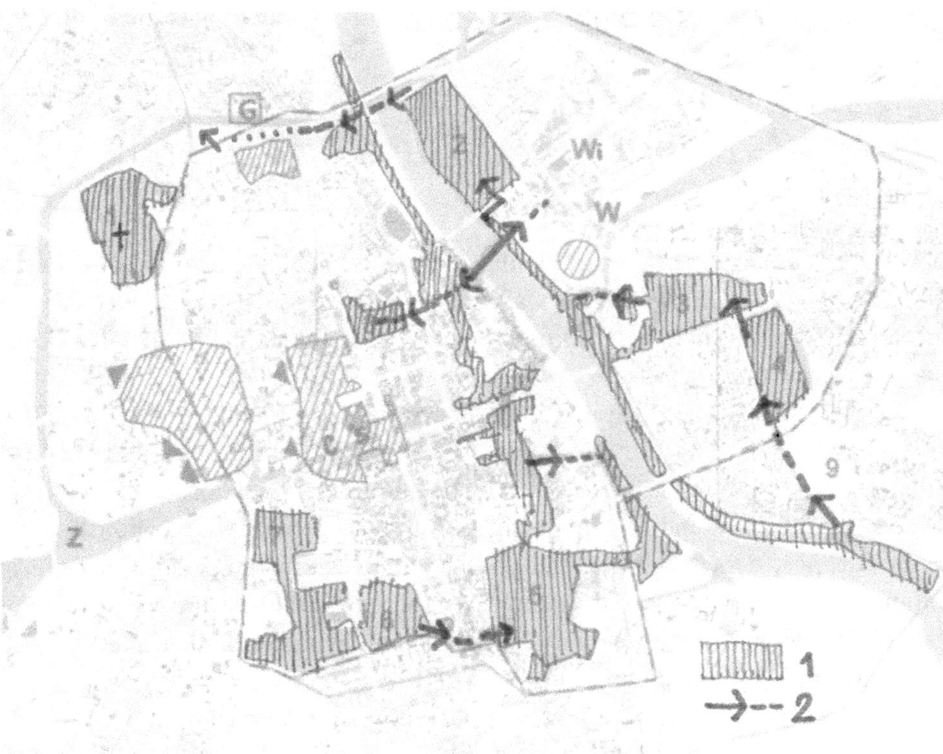

Figure 14. A fragment of the communication route Świętokrzyska Street after reconstruction, introduction of greenery, bicycle paths and narrowing of the carriageway. The investment has been widely praised
(photo by the author, 2018)

makes it impossible to plant trees with broad roots. Such streets full of greenery attract people who are willing to walk there. If there are service outlets on the ground floors of neighboring facilities, they gain new clients and financial turnover increases along these streets.

An interesting idea, which is gaining in popularity in Poland, are the so-called social gardens. The first activities of this type have been undertaken and they met with positive response, so there is hope that the implementation of this concept will happen quickly and to the satisfaction of participants. Large Polish cities are increasingly deprived of greenery. Hence, the need to create gardens by local communities. The purpose of this activity is not only to raise the aesthetic value of residential areas, but also to integrate local communities and strengthen interpersonal relations in neighborhood units.

Dominika Kulczyk,[11] the founder of the Green Cross Poland association, also runs projects in the field of corporate social responsibility, social activities dedicated to local communities in Poland and abroad, charitable and ecological actions. Green Cross Poland promotes voluntary service in developing countries. The association promotes and supports e.g. functioning of social gardens, which have been created in Polish cities since 2015. This is usually done by assigning a moderator who instructs all participants on planning, sowing, collecting crops and their use.

The crucial element of this project is the possibility of obtaining grants for creating social gardens. Neighborhood groups established for the purpose of participating in this project may apply for various types of support for active gardening and reinforcing real dialogue in neighborhood communities. In today's world of digitization, IT networks and symbiosis with machines (telephones, computers, etc.) such interpersonal relations based on dialogue are invaluable.

The purpose of creating the Urban Greenery Board of Warsaw is systematic and comprehensive management of the urban environment, especially created by greenery and water, including the Vistula embankments. It was decided that the governing principle of its activities would be the creation of a continuous system of interconnections, with green streets and existing greenery enclaves - a network system characteristic of modern metropolitan cities.

In this context, the concept of creating "rain gardens" has emerged, which would be located mainly within the former Russian fortifications from the end of the 19th century. These sites are difficult to develop due to their large size, the fact that they encompass green areas and buildings, usually partially covered with soil, e.g. Fort Czerniakowski, where the concept will be first implemented. The project provides for the installations with vegetation species using rainwater, which, therefore, be used on the spot and not discharged into the rainwater drainage system. The project also involves development of a rainwater management system for streets.

It is important to note that there are initiatives to increase the number of children's playgrounds in the city, both in public spaces and in residential areas, whereby empty courtyards or other common spaces are revived. It is assumed that these playgrounds must have educational value and ecological form. Considering the quest for natural greenery of native character, the trend of introducing meadows with natural vegetation in the city will be widened.

One of the most important recent linear designs in Warsaw are the Vistula Boulevards, where walking and bicycle paths are accompanied by attractive cultural and entertainment offering for the users of this recreational area. This investment is of course appreciated as a valuable development of the river embankments. the ecological arrangement will not include concrete surfaces; pedestrian trails will be unpaved.

Such arrangement of these spaces should encourage people to spend more time there and, as a consequence, lead to the economic revival of the Vistula River embankments.

CONCLUSION: POSSIBLE CHANGES IN SPATIAL MANAGEMENT

For a long time in the second half of the 20th century, Warsaw and its city center were an ecological city with a lot of greenery, not very dense building development. The city destroyed by the Germans during World War II was gradually rebuilt. It had a characteristic structure with green wedges converging towards the center - a model urban layout. After the political change in 1989, joining the European Union and the implementation of democratic principles, several changes have taken place: the compliance with earlier urban regulations was lessened, developers and, currently, politicians have been given greater freedom in decision-making. Green ventilating wedges have been built-up, with the exception of the areas located in the immediate vicinity of the Vistula River, the intensity of building developments, especially in the city center, has increased, and apparently unlawful felling of all trees in the city was allowed, although it made by official state services.[12]

These controversial actions caused social protests against the authorities dealing with environmental protection, making disadvantageous decisions for the climate in the city and the health of the residents. In the autumn and winter season, blue-gray clouds of smog pollute the air in Warsaw. This state caused the necessity of positive actions of municipal institutions for the ecological improvement city, to which they were forced. As a result, the following activities have been selected and gradually implemented.

1. Deepening cooperation between scientists, academic institutions of Warsaw and employees of municipal offices, in order to give officials knowledge about current, proven theories of urban and environmental development and practical principles of transforming and revitalizing urban space.
2. Organization of courses, meetings, conferences and workshops with the participation of many groups of stakeholders from the city center and other parts of Warsaw, exchange of views, publications, presentations of opinions.
3. Care of historic and contemporary parks and squares of Warsaw, taking into account their biological values for people and the use of economic opportunities for the development of these places. Broadening the range of functions that

accompany these parks: exhibition, educational, health-oriented, catering, etc., in order to increase the number of users of these areas.

4. Research on the potential of the Natura 2000 area on the right bank of the Vistula River as an element of bioeconomy.
5. Creating a network system from the space of parks and public squares in the area of the city centre of Warsaw. Indication of missing links in this system and examples of solutions for the sections of urban streets connecting larger greenery complexes.
6. Connecting the created greenery system in the city center with selected service functions to stimulate the principle of bioeconomy.
7. Transform the existing poorly developed areas in the center into one coherent system of links by walking and bicycle paths, other ecological means of transport (e.g. electric cars); it should create the possibility of moving to various interesting areas of the city center, worthy of spending time there.

The analysis in this chapter of the existing natural habitats and developed biological forms of the city centre of Warsaw shows the development potential of these elements. However, this requires thorough research, observation of the results of individual actions and, above all, wise, responsible management in the scale of larger areas.

REFERENCES

Chmielewski, J. M. (2001). *Teoria urbanistyki w projektowaniu i planowaniu miast*. Warszawa: Oficyna Wydawnicza Politechniki Warszawskiej.

Encyklopedia Warszawy. (1994). Warszawa: Wydawnictwo Naukowe PWN.

Guranowska-Gruszecka, K. (2013). *Śródmieście Warszawy w XX wieku, Warszawa wyd. Szkoła Wyższa im*. Bogdana Janskiego.

Guranowska-Gruszecka, K. (2017), *Shaping of the Centre of Warsaw –history of activities and views, visions of future*. Szczecin wyd. PAN DOI: 1021005/pif.2017.30.A-01

Łaskarzewska, M. (2015). Piękno i użyteczność "Polskiej Zielonej Ściany". *Zeszyty naukowe uczelni Vistula, 42*(4), 60 – 69.

Pałac Saski w Warszawie zostanie odbudowany? (n.d.). Retrieved from http://historia.org.pl/2011/05/06/palac-saski-w-warszawie-zostanie-odbudowany/

Stefańska & Wieczorek. (2018). *Diagnoza użytkowania Pola Mokotowskiego*. Raport z badań psychologii Środowiskowej.

Wiesław. (n.d.). *Warszawskie pomniki*. Warszawa: Wydawnictwo "PTTK Kraj."

Yung-Zielonko K. (2012). *Łączenie zaawansowanych i tradycyjnych technologii w architekturze proekologicznej"*. Oficyna Wydawnicza Politechniki Warszawskiej.

Zuziak, Z. (2001). Strategies for Large Scale Urban Developments – the Game between the Public and Private Sector. In *Large Scale Urban Developments*. Academic Press.

Zuziak, Z.K. (2004). *Strategiczny wymiar urbanistyki*. Kraków: Politechniki Krakowskiej.

KEY TERMS AND DEFINITIONS

Center of Warsaw: The central borough of the city of Warsaw.

Green Urban Network: System of parks, green open spaces and public squares in the area of the city.

Green Ventilating Wedges: Green open spaces that allow the flow of fresh air.

Greening Strategy: A strategy for improving the health of residents; the investments planned for implementation until 2020 set the directions for improving the living conditions in Warsaw.

Natura 2000: Areas belonging to the network of protected natural areas in the territory of the European Union, in order to preserve specific habitats and valuable species of wildlife.

Traffic Generators: Places which attract the largest number of space users; attractive spaces that encourage people to spend time in them and use all facilities, innovations, educational elements, opportunities for recreation and practicing various forms of sport.

Warsaw: Capital of Poland, among 22 cities in other countries, is already perceived as the world's metropolis.

ENDNOTES

1 In 1989, political transformation took place in Poland; it meant moving away from communism, building a free market, creating a civil society and democracy.

2 The radial system has been applied since 1916, when Poland regained independence after the partition (Warsaw was under Russian rule). The radial system was fully defined and implemented from 1931, when it was indicated in the regulatory plan of the capital city of Warsaw

3 After the destruction of the Berlin Wall and the eastern and western connections, Germany has a surface area slightly higher than Poland, 357,000 km2. In contrast, countries such as Finland, Sweden, France or Spain are powerfully stronger.

4 In 1940, the Germans blew up the Chopin monument. The statue weighing 16 tons was cut to pieces, taken to the German Reich and mostly used for the production of weapons [https://www.polskieradio.pl/39/156/Artykul/854658,Lazienki-no -Chopina? ...]

5 The Mokotow Field is located between the following streets: Żwirki i Wigury, Rostafińskia, Rokitnicka, Batorego and Niepodległości Av.

6 http://metro.warszawa.gazeta.pl/mettrowatszawa/7,141637,23645833,..." Niemcy wychwalają Warszawę. To najbardziej tętniące życiem miasto w Europie [access: 2018-7-10]

7 The axis runs from Jazdy Polskiej Roundabout through the footbridge next to the National Library, to the intersection of Żwirki i Wigury St. and Banach St.

8 Permaculture is a response to threats resulting from modern intensive agricultural crops, which threaten the depletion of natural resources and environmental degradation. Permaculture adheres to three ethical principles: concern for the land, for people and for the return of excess. The permaculture garden is designed in such a way that it functions in harmony with nature.

9 The Urban Greenery Board was established in 2016 with a new team of young, talented and dedicated people, which brings positive results in the form of implementation of green corridors in the city.

10 Pluta; Zuziak

11 D. Kulczyk, born 1977 in Poznań, founder and president of the Kulczyk Foundation, founder and vice president of Green Cross Poland, she is also a member of the Supervisory Board of Kulczyk Inwestments, President of the Board of the Polish-Chinese Business Council, co-founder of the group of consulting companies Values.

12 For example felled trees in front of the City Hall in the centre of Warsaw (2017) during the weekend.

Chapter 7
Green Spaces of the Metropolitan Area of Guadalajara

Roberto Jiménez
Universidad de Guadalajara, Mexico

Paula Lourdes Guerrero Rodríguez
Universidad De Guadalajara, Mexico

Rogelio Rivera Fernández
Universidad De Guadalajara, Mexico

ABSTRACT

The analysis of some systems of green areas and public parks of the metropolitan area of Guadalajara, other cities of our country Mexico and Latin America, shows common problems such as the deficit of urban green spaces, insecurity, unemployment, and uncertainty with a social exclusion in these areas of stress. Likewise, the lack of economic value of the services provided by such natural systems as recreation is added. Together they are important factors in the allocation of territories destined to this use with respect to others that generate Urban speculation. Therefore, it is proposed to develop a typology of green areas appropriate to the needs of the metropolitan region. It will facilitate the production of inventories that estimate indicators of territorial cohesion, governance, economic profitability, social, environmental quality and innovation, as well as incorporating new technologies that improve geographic information systems and internet media that support management.

DOI: 10.4018/978-1-5225-7958-8.ch007

INTRODUCTION

In the planning of urban green spaces, citizen participation is relevant, as a basic strategy to solve the difficult situations that are presented in the ZMG; Must ensure inclusive processes, especially to cover demands for the spread of vulnerable social groups. Paquot Thierry (2016).

It is recommended to calculate the bio-Economic values related to recreation through the Model of the Quintuple Helix model with the support of quantitative, qualitative and comparative methods. It is concluded that it is necessary to approach green planning based on the philosophy of the system of indicators that above all must be a real and applicable system that measures the degree of sustainability of urban green spaces in the ZMG. Anaya (2008).

The Quintuple Helix innovation model is even broader and more comprehensive by contextualizing the Quadruple Helix and by additionally adding the helix (and perspective) of the 'natural environments of society'. The Triple Helix acknowledges explicitly the importance of higher education for innovation. However, in one line of interpretation it could be argued that the Triple Helix places the emphasis on knowledge production and innovation in the economy so it is compatible with the *knowledge economy*.

The Quadruple Helix already encourages the perspective of the *knowledge society*, and of *knowledge democracy* for knowledge production and innovation. In a Quadruple Helix understanding, the sustainable development of a knowledge economy requires a co-evolution with the knowledge society. The Quintuple Helix stresses the necessary *socio-ecological transition* of society and economy in the twenty-first century; therefore, the Quintuple Helix is ecologically sensitive. Within the framework of the Quintuple Helix innovation model, the natural environments of society and the economy also should be seen as drivers for knowledge production and innovation, therefore defining opportunities for the knowledge economy.

The European Commission in 2009 identified the socio-ecological transition as a major challenge for the future roadmap of development. *The Quintuple Helix supports here the formation of a win-win situation between ecology, knowledge and innovation, creating synergies between economy, society, and democracy*. Global warming represents an area of ecological concern, to which the Quintuple Helix innovation model can be applied with greater potential.

Therefore, there are nine areas, of which Carayannis and Kaloudis write about, that require 'sustained action', political and economical 'leadership' or 'empowerment', and 'intelligent use of technology' (Carayannis and Kaloudis [2010], p. 2):

1. “Financial/economic system;
2. Environmental challenges;
3. Feed and heal the world challenges,
4. Energy challenges,
5. Educational challenges,
6. Political democratic reform across the world,
7. Transformative government across the world,
8. Equity and Security across the world,
9. Technology, innovation and entrepreneurship as drivers of knowledge societies”.

Let us consider now in greater detail the production of the resource of knowledge. Knowledge (for example, the advancement of green technology) can act as key to success for sustainable development. Essentially, it should be understood today that nation-states that concentrate on the progress of society, higher competitiveness of their economies, or better and sustainable quality of life have to apply the resource of knowledge. In the transformation to a knowledge-based society, knowledge-based economy, or knowledge-based democracy (see Carayannis and Campbell).

Urban wooded reduce global warming, or “the green house” effect in the cities (figure 1). Produced by the excessive warming due to lack of green areas, air contamination and the big heat reflecting surfaces, as pavement and concrete. Regulation of parks and gardens of Guadalajara (2014).

In Figure 1 presents the greenhouse effect and Solar Radiation.

Green Areas Benefits

- Produces oxygen.
- Reduces noise level.
- Regulates weather.
- It becomes an ecosystem for birds and pollinator insects.
- Embellishes the urban landscape.
- It changes the city noises for nature sounds.

Aims

- To contribute in the fulfillment of the paradigms mentioned in the model of the fivefold helix with the coolness of the 5 inclusive spheres in benefit of the rational use of the spaces in the ZMG.

Figure 1. The greenhouse effect and Solar Radiation

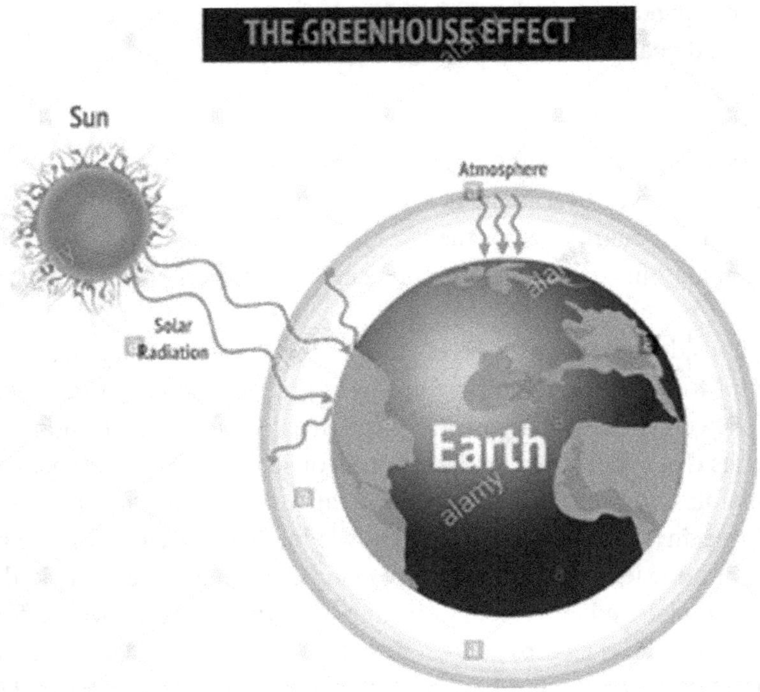

- Set natural green areas, parks and public spaces in the Guadalajara Metropolitan area.
- Propound decent spaces of recreational use.
- Indicate suitable guidelines on environmental politics.
- Suggest a new healthy lifestyle culture.

METHODOLOGY

The present study is of correlative depth since it delimits the important correlation that must exist between the fundamental elements mentioned in the model of the fivefold helix with support also with the general theory of system among these, and the participation of the government, businessmen, universities, society and the environment.

It has been complemented with a methodology of methodological triangulation, in which theorists, data, techniques, tools and data sources are intertwined, especially with a pragmatic approach.

It starts with guiding questions and what information should be consulted is defined. It forces to review the existing quantitative and qualitative information.

In the research methodology the public spaces of the Guadalajara Metropolitan Area, the quintuple helix model is considered, conceived in its dawn by Henry Etzkowitz (2000), when considering the link between governmental task, the development of the companies, the task performed by the higher education institutions.

It is complemented by the participation of society, and the environmental balance.

Some experts recommend urban living labs, but in Mexico we don´t have this kind or resources taking the environment as urban living lab. To continuing to describe the five basics elements that is configured the theory of the quintuple helix.

1. Government
2. Enterprises
3. Universities Bautista, E. (2015)
4. Society
5. Environment

Variables

- Lack of planning.
- Speculation of urban area.
- Lack of environmental education.
- Authorities'' apathy.
- Business people's economic interests.

Hypothesis

In the case of not implementing the model of the quintuple helix in green areas, they will get extinct in 2050.

Health Benefits

- According to the World Health Organization, a minimum of 12.5 square meters of green areas per inhabitant within the city is recommended, to contribute to the quality of life of the people.
- This contributes to decrease chronic respiratory diseases in children and the elderly mainly.

- The presence of vegetation blocks solar radiation, which is harmful for people's health. Agenda Habitat, Villasis Keever, R (2005).

Environmental Deficit

One of the main problems is that there are more vehicles per square meter than trees, which implies an exaggerated emission of pollutant particles.

The environmental emission is: 1'000,000 of trees with the presence of 2'000,000 of vehicles. Equivalent to 1 trees for every 2 vehicles. Agenda Habitat, Villasis Keever, R (2005).

Public Green Areas Estimation in the Municipality of Guadalajara

Guadalajara city currently presents environmental problems caused by several factors, among them the insufficiency of green areas (Figure 2) that is multifactorial, the above was articulated to a lack of city's growth planning, since it was not contemplated the conditioning and maintenance of established areas, neither the incorporation of reserves or green areas.

It was obtained from each of the seven areas (Centro, Minerva, Huentitán, Oblatos, Olímpica, Tetlán and Cruz del Sur) the total surface as well as the public green areas (central reservations, fountains, gardens, roundabouts, parks, squares, triangles); and the population census of each of them. Regulation of parks and gardens of Guadalajara (2014)

In Figure 2 presents The deficit of green areas in the Guadalajara Metropolitan.

According to the Institute of Territorial Information (IITEJ for Its Acronym in Spanish)

In the final columns of the chart, it is quantified the average quantity of square meters per inhabitant per geographic area (Table 1).

Square meters of green areas with vegetation per inhabitant is indicated in Table 1.

It can be seen how the worst profile of geographic green areas is the one of Guadalajara, because it has less than 4 m^2 of green areas per inhabitant.

Guadalajara has an average of 21.34 m^2/inhabitant, this is an atypical case since in this municipality we can find the Huentitán canyon, because of that the average has been represented that way (Table 2).

Figure 2. The deficit of green areas in the Guadalajara Metropolitan

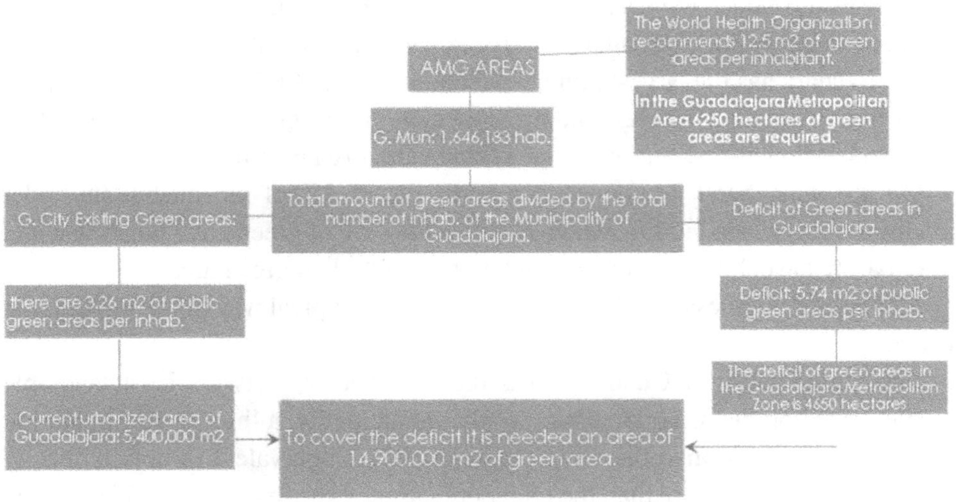

Table 1. Square meters of green areas with vegetation per inhabitant.

		Square meters of green areas with vegetation per inhabitant				
		Critic	Severe	Very Low	Low	Optimum
	Population	0 a 2 m²	2 a 4 m²	4 a 6 m²	6 a 9 m²	More than 9m²
Guadalajara	1`494,309	15.77%	18.03 %	13.50 %	14.81%	37.90%
Tlaquepaque	608,686	10.37 %	7.26 %	9.55 %	9.02 %	63.80 %
Source: Territorial Information Institute from image Spot 2009 and population data from INEGI, 2010						

Table 2. Percentage of population per municipality according to the quantity of square meters of vegetation per inhabitant.

		Green areas average per inhabitant						
		Critic	Severe	Very Low	Low	Optimum		
	Urban Areas	0 a 2 m²	2 a 4 m²	4 a 6 m²	6 a 9 m²	More than 9 m²	Median	Average
Guadalajara	982	14.56 %	16.90 %	14.05 %	15.23 %	39.21 %	6.50 %	21.34 %
Tlaquepaque	199	15.08 %	10.55 %	11.56 %	12.56 %	50.25 %	7.08 %	59.41 %
Source: Territorial Information Institute from image Spot 2009,								

If the amount of inhabitants that live in each urban area according to the quantity of vegetation is added to chart 1, the outcome would be chart 2.

Green areas average per inhabitant is indicated in Table 2.

Guadalajara has the largest number of inhabitants -> 61% with less than 9m^2 of green area/ inhabitant. Corresponding to them 4.5 m^2 of green area per/ inhabitant being well below the mean set by the world health organization.

On the other hand Tlaquepaque records the 36.3% of its inhabitants under conditions of low availability, which is below the 9 m^2 of green area/ inhabitant and these barely have 3.8 m^2 green area/ inhabitant of "urban green areas".

Inhabitants with less than 9m. ^{2}/ inhab. There is a typical way of life of Mexican leader.

The municipality of Guadalajara is the one that presents the least favorable profile, based on the total population of the municipality, a third of its population does not reach more than the 4m^2/inhabitant, this is equivalent to approximately 504,939 inhabitants, more than the total population of the municipality Tlajomulco or Tonalá. Web site:

Public Spaces and Recreation Areas in the Guadalajara Metropolitan Area (ZMG) is indicated in (Table 3).

Open and recreation spaces affirm the requirement of the relationship of society constituted by several elements, which are generated through objectives, functions

Table 3. Public Spaces and Recreation Areas in the Guadalajara Metropolitan Area (ZMG)

ÀMG	Guadalajara	Zapopan	Tlaquepaque	Tonalá
Natural Protected Area	2	3	-	-
Park	24	98	15	39
Garden	250	136	20	27
Sport Complex	87	34	26	27
Squares	37	9	11	11

and the respective transformations (Anaya 2008). Venues of expression and urban appropriation, administered by public organisms. Each day there is a less green garden in Guadalajara.

Research Question: What Do These Spaces Favor and Contribute to?

They contribute and favor to an environmental education, comfort, sport, recreational, and cultural activities and if it was not enough recess, increasing the economic value of the properties How? Adding surplus value. Author (2017)

Open Spaces of Public Use in the Guadalajara Metropolitan Area

A total of 851 public open spaces and 5 natural protected areas were registered in the Guadalajara Metropolitan Area.

It is important to know that in the four city councils, mark all the spaces that are administered by "parks and gardens" as "green areas".

Figure 3 presents Public open spaces and natural protected areas in the Guadalajara Metropolitan area (2017).

Area of Influence or Green Spaces Service

- Neighbors: 300 - 2500 m^2.
- Neighborhoods: 2500 - 4600 m^2.
- District: 4600 – 10,000 m^2.
- Central: 10,000 – 10,000 m^2.
- Regional: 10 – 100 ha.
- Natural Protected Areas: more then 100 ha.

Territory

It is estimated that in the year 2030 world population will reach 8.5 billion people, leading to degradation of the environment, in such a situation territorial size is an indicator to label a green and recreational space, this from the point of view of urban and environmental classification. Oberti Marco et Préteceille (2016).

One of the standards to consider is the minimum area for a space to reach the recreational or "green" space category, an example is the German standard that considers surfaces larger than 17 m^2 per person.

Figure 3. Public Spaces and Recreation Areas in the Guadalajara Metropolitan Area (ZMG)

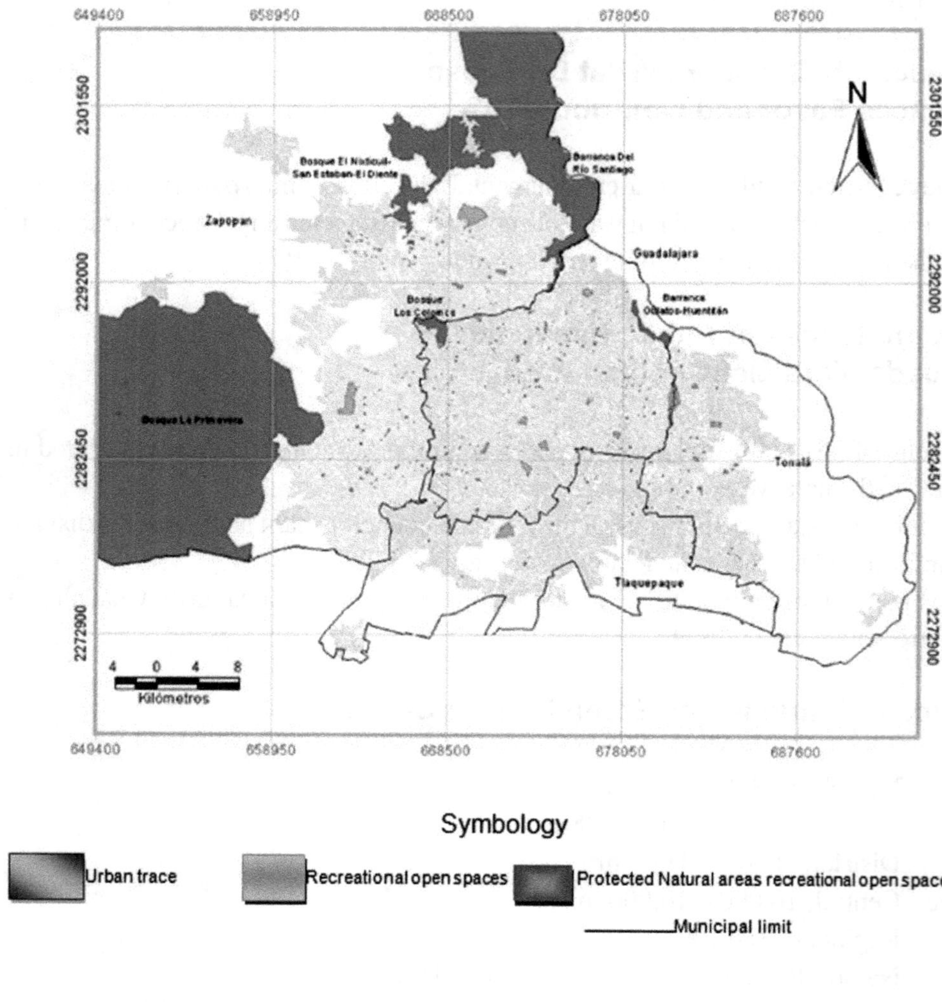

This indicator, unfortunately, has not been fulfilled, due to the high political and economic interests that have had throughout its history in our country. In such a situation, it is desirable to apply the fivefold helix in every way to counteract the contaminating and degrading effects of the environment.

In the case of Mexico and specifically in the Guadalajara Metropolitan area the following distribution of recreation areas is discussed. Citizen council metropolitan Guadalajara (2016).

ListofPopulation and open recreational green areas in the Guadalajara Metropolitan Area (2005) is indicated in (Table 4).

Table 4. Population and open recreational green areas in the Guadalajara Metropolitan Area (2005)

List of population and open recreational green areas in the Guadalajara Metropolitan Area (2005)					
Municipality.	Percentage of metropolitan population.	Population	Open recreational areas.	Open recreational areas. (m^2)	Surface area m^2/inhab Open recreational area
Guadalajara	49.42	1600940	398	4061758,96	[illegible],53
Tlaquepaque	13,77	563006	72	851083,00	[illegible],51
Tonalá	9.47	408729	104	1171952,28	2,87
Zapopan	27,34	1155790	277	3995858,89	3,46
AMG	100%	3728465	851	10080653,00	2,70
Source: Own elaboration based on basic tabulation of Population and housing , INEGI, 2005 and the open recreational areas system generated.					

It can be observed in the chart that the municipality of Guadalajara has the largest number of open spaces, Zapopan that occupies the second place, presents a greater area of m^2 of open spaces, that is to say that Zapopan owns less spaces but they have greater area with an influence on Lake Chapala (Figure 4).

Reforestation proposal at Chapala´s shore lake: towns bordering the Chapala´s lake ins indicated in (Figure 5).

Figure 4. Chapala's Lake

Figure 5. Reforestation Proposal of Chapala´s Shore lake: Towns Bordering the Chapala´s Lake

Proposal solutions: Guadalajara Metropolitan Area and Public Spaces Nearby is indicated in (Figure 6)

In relation to the study area of the Metropolitan Zone of Guadalajara (ZMG), an assessment was made applying situational analysis considering four factors strengths, opportunities, weaknesses and threats of the public natural spaces surrounding the area with scarce tourist use, which together represent more than 100 thousand hectares, diagnostic of the internal situation reflects a weak contribution to Guadalajara to be considered a sustainable city, also its external evaluation was similar, meaning professors undertake to get better sustainable condition in the University of Guadalajara. It is an introspective pre-diagnostic exercise, where the identification is made through a previous analysis, with the greatest possible objectivity, of the characteristics of the government, university and company that make it strong in the face of competitors.

Figure 6. Proposal solutions: Guadalajara Metropolitan Area and Public Space Nearby

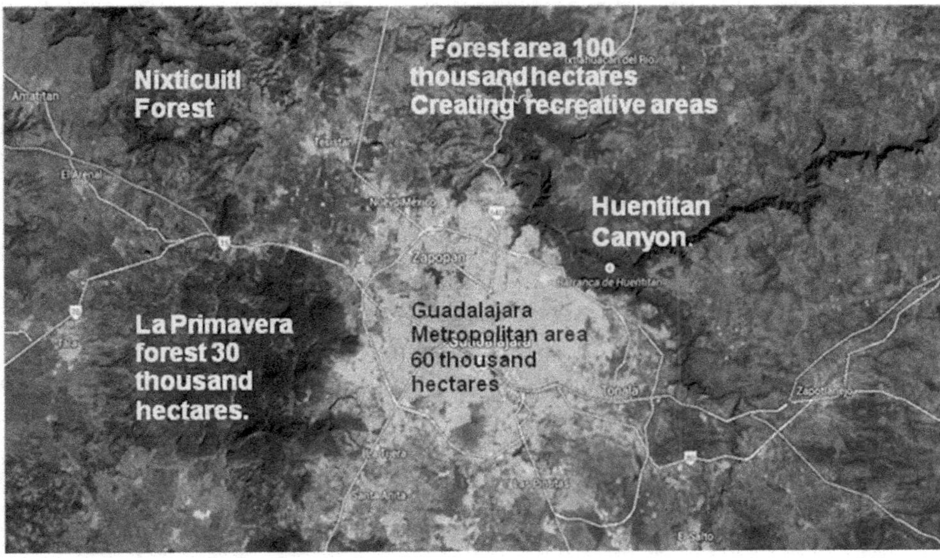

These concepts are applied in the integral studies on the viability of an entrepreneurship project, through which the situation and real possibilities of a restricted economic activity in that rural portion adjacent to the ZMG are valued.

The valuable natural heritage that the region presents, we must add the interesting cultural heritage, since in that area there are pre-Hispanic vestiges that have not yet been rescued and the existence of old hacienda halls producing tequila, agricultural and sugar mills, dating from the eighteenth century to the nineteenth century. And so the municipalities contemplate making the creation of museum spaces and a ball game as practiced by the Aztecs.

Also, the orographic characteristics of the area and the existence of a low population density (20 inhabitants per hectare) are circumstances that favor astronomical observation and contemplate the interesting flora and fauna resources. With all this, a new proposal arises that is currently being evaluated, incorporating a natural space of more than 100 thousand hectares surrounding the ZMG. Initiatives and decisions in the adjacent area can converge on points of tourist attraction and wide-ranging recreation for the local population, contribute to generate economic resources in the area and give a different image to that in the present. In this virtue, it becomes in this way a real opportunity for the ZMG and its surrounding rural environment.

The Forest of Primavera close to Guadalajara is a fraction of the transverse neovolcanic axis. The origin of this mountain range is very recent and begins when volcanic eruptions occurred (between 140,000 and 27,000 years ago), during which time a series of estimated pyroclastic flows in the order of 20 km.2 were emitted, covering an approximate area 700 km2 distributed mostly over what is now the ZMG; When this event occurred, a superficial part of the Earth's crust was caused to collapse causing a depression that originated a lake, currently extinct, of a diameter of approximately 11 kilometers; in it, lacustrine and volcano-sedimentary sediments were deposited, over which, or interspersed with them, some lava flows (rhyolite) flowed. As a remnant of this volcanic activity, a hydrothermal activity has remained in the area that manifests as fumaroles and springs of hot water. The region represents a geothermal potential used for tourism purposes of natural thermal manifestations, and the convenience of generating electricity through geothermal fields is under study.

EVALUATION OF THE AREAS WITH TOURIST INTEREST IN THE GREAT CANYON OF THE OBLATOS-HUENTITÁN

The Canyon should be considered as one of the main natural attractions of the Municipality, not only because of the impressive view it presents, but also because of the micro climatic variety, which favors a great biodiversity. Its cultural importance must also be mentioned, given that it was used as a passage in the Guadalajara-Zacatecas trade route, carried out by muleteers for several hundred years.

The protection of the Great Canyon of the Oblatos-Huentitán awakened interest in the seventies, but it was not until the last dates that there was interest from environmental groups to make this place a recreational center and it became a public concern, as it was threatened by changes in land use and loss of biodiversity (Figure 7).

Another problem is the scarce tourist value that has been given, and the lack of infrastructure to enjoy it, since it only has:

Two viewpoints created in the 1970s, the "Independence" at the end of the Independence road and the "Gerardo Murillo, Dr. Atl, at the end of Belisario Domínguez Avenue, where there are food areas, playgrounds, and courts Sports.

Guadalajara Zoo: its construction stands out as attractive, the natural unevenness of the terrain due to its proximity to the Canyon, having at its limit a viewpoint to admire the impressive view.

Figure 7. Huentitán Canyon

Parks of Oblatos: 210 hectares where families make field days, which is reached by the East Peripheral Ring, which offers attractive natural landscapes. It is also used for sports activities, since there is an athletic track, in addition to recreational and recreational activities. There is a hot springs spa, with quite rustic facilities, and unfortunately about to fall by carelessness of the authorities.

Use of Public Spaces of Urban Forests Within the ZMG

In the history of cities, it is perceived how, through their growth and development, the different urban green areas are transformed. These are divided into open or public spaces, recreational and leisure places (such as parks), and restricted or private access sites, which are accompanied by dwellings where the green areas are indicative of a social status. Law Official Gazette of the Mexican Federation (DOF, 1988).

The issue of ecological balance in the metropolis has gained greater importance in the international arena since the First Summit Conference of the United Nations Organization on the Environment, in the city of Stockholm, in 1972 (Gonzalez, 1996).

The direct antecedents on the subject date from 1976 with the summit conference on the cities "Habitat I", in Vancouver, Canada. Subsequently, in Istanbul, Turkey, in June 1996, the second summit conference of the cities of the United Nations Organization on Human Settlements "Habitat II" was held. The central theme: the problems generated by the accelerated development of cities.

In Mexico the Institutions That Have Worked on the Studies of the Urban Ecology Are

The Institute of Ecology, AC, the Faculty of Sciences of the National Autonomous University of Mexico, the Azcapotzalco Unit and Iztapalapa of the Autonomous Metropolitan University, the Department of Studies Urban and Environmental Affairs of The College of Research of the Northern Border, the Mexican Society of Natural History and the University of Guadalajara, among others; Research generally considers urban parks within the green areas of a city, either as components of the internal structure of the city or as functional elements of the urban, from which the following these approaches can be distinguished for their study:

First, the ecological approach. Where they study, within the gardens and wooded areas of the park, species of flora and fauna through inventories, as habitat, and the existing relationships between them. In addition to considering urban parks in other aspects: its green areas influence the microclimate and the recharge of aquifers, its wooded areas contribute as curtains breaks wind that reduces dust, noise and unpleasant aromas.

Second, the approach to landscape architecture. Visualize urban parks from the perspective of architecture, that is, consider the aesthetics and operation to improve the appearance of a place for users. Consider the part of the built equipment, sculptures, monuments, gardens and wooded areas

Third, tourism focus. In these studies the dates of creation of the park, the objectives they pursue, who inaugurated it, where it is located, what is its importance with respect to the traditions of the site, services and attractions offered to visitors.

The above approaches show us that parks, although in the State Law of Ecological Equilibrium and Environmental Protection (1989) are considered as a category within protected natural areas, have generally been absent from studies towards planning. Under these premises, the urban park is an open space for public use. It establishes human relationships of leisure, recreation, sports, community coexistence, education and culture within the city. Express in the concrete one of the forms of relationship society-nature. On the one hand, they are being governed by laws of a biological nature and, on the other hand, they have a social function. Such recreational spaces are part of the green areas in the urban context. They are built with a social objective and acquire different connotations, there being several factors that determine them: physiographic, sociopolitical and cultural.

It is important to point out that these types of recreational green areas have a social essence, since society assigns it, and they contain an objective value. At the same time they are valued subjectively in correspondence with the individual interests of each person. In this way, the value becomes determined by the objective significance that they represent in their social essence, while the valuation has a subjective character, and it is difficult to specify the meaning of the urban parks for each individual, since it varies according to the to be different the needs, aspirations and satisfactions of each one.

Urban Parks for Public Use Within the ZMG

Conceived as open spaces for public use, it establishes human relations of leisure, recreation, sports, community coexistence, education and culture within the city. Express in the concrete one of the forms of relationship society-nature. On the one hand, they are being governed by laws of a biological nature and, on the other hand, they have a social function. "These spaces integrate the human being and knowledge of a specific historical epoch" (Miranda, 1997).

It is observed that the current urban parks for public use within the ZMG, mostly manifest a series of deficiencies and basic needs related with the maintenance and improvement of infrastructure and these are reflected in the signage of the infrastructure, need to rehabilitate areas of tourist interest, poor accessibility, there is a degradation in the spaces close to the population centers, safety on the route, poor quality due to the deterioration of the floors, flooding areas, equipment damaged in the rest areas, others.

The parks in the urban environment of the ZMG are the result of the practical activity of the local population, since they contain a natural component (flora and fauna) and a cultural partner one that reflects the world view, customs and traditions of society.

It is important to note that these types of recreational green areas have a social essence, since society assigns it, and they contain an objective "value". At the same time, they are valued subjectively in correspondence with the individual interests of each person, according to the values and functions.

It is evident that urban parks are necessary spaces for the improvement in the quality of life of the citizens of urban areas. Despite this, many of the time they are underutilized due to lack of planning.

The above is reflected in the ZMG, where the most representative parks and natural areas are irregularly distributed with a figure of 756 and of which only five of them (Agua Azul, Oblatos-Huentitán Canyon, Los Colomos Forest), Metropolitan and Bosque El Centinela) environmental education activities are carried out as part of the services offered to visitors and school groups.

Although these spaces represent an enormous complementary or parallel potential to higher education in the field of environmental education and related subjects, they have not yet been open to professional research and practice. Hence the need to conduct research systematizing information about them, and disseminate to the public in order to know them. Because it has been identified a lack of knowledge on the part of the inhabitants of the ZMG of the different existing recreation options.

Also, involve governmental, non-governmental institutions and the functions of the parks in the urban environment.

It should be noted that the dimension of Natural Resources and Environmental Protection, includes two major concepts: on the one hand the availability of natural resources that Jalisco has and on the other the protection provided to said. Likewise, the city of Guadalajara currently presents environmental problems caused by several factors, among them the insufficiency of green areas that is multiple research factors, the previous one was articulated to a lacking planning of the growth of the city, it was not contemplated the conditioning and maintenance of areas established, nor the incorporation of reserves or green areas. It was obtained from each of the seven zones (Centro, Minerva, Huentitán, Oblatos, Olímpica, Tetlán and Cruz del Sur) the total surface as well as the public green areas (ridges, fountains, gardens, gazebos, parks, squares, triangles) and the population census of each of them. (Municipal Presidency of Guadalajara, 2017).

The transformations in the behaviors that can be fostered and appropriated by the civil society organizations and citizens are crucial for there to be a positive change in the face of the great challenge to have more than 9 square meters per inhabitant that will contribute to climate change of quality in the surroundings of Guadalajara. Our education and vision regarding the future, have to adapt to new parameters. In such a virtue, the prospective look of a "green" architecture that is concerned with preserving the former City of Roses is crucial, to ventilate naturally, with the reuse of rainwater, solar facades or devices for the use of wind energy. Being complementary, the incorporation of green roofs with the presence of vegetables that contribute to the economic development of the region in the food line.

More than in any other city in the state of Jalisco, people often ask themselves if the air quality has improved or worsened. Perhaps due to the lack of credibility in the government programs, the occasional application of the contingency plan and the daily evidence of not seeing the blue sky, there is a perception in many people that air pollution is greater and that the measures and control programs applied in recent decades. However, apart from demagogic speeches, in 2016 there has been an improvement in air quality. For example, even at the beginning of the 21st century, the six pollutants measured in the automatic monitoring network often exceeded the corresponding air quality standards. Anaya, C. M. (2008). And (SEMADET, 2016).

However, regardless of demagogic triumphalism, in the last decade there has been an improvement in air quality. For example, even in the early 2010´s, the six pollutants measured in the automatic monitoring network often exceeded the corresponding air quality standards.

In the ZMG where high levels of pollutants are recorded by heavy metals or aromatic hydrocarbons, the number of people affected by diseases such as leukemia, lymphomas, cancer and even personality disorders, such as depression and over excitation, has increased. “Intense contamination has begun to be a risk factor”, recognize several researchers from the Division of Biological and Environmental Sciences of the University Center of Biological and Agricultural Sciences of the University of Guadalajara.

Likewise, a group of experts from other Universities and Research Centers in the field of neurotoxicology has investigated the impact of contaminants and it is detected that patients with leukemias and lymphomas have begun to appear at younger and younger ages. including in childhood and adolescence. The same happens with breast cancer, which previously affected women aged 45 to 55 years, on average, and today is common in young people of 25 years.

In addition, the lack of a connectivity plan between the different places of habitation, work zones, schools and service areas is detected. The situation worsens especially in the eastern area in the Oblate colony and south in the Miravalle colony, having places with high rates of air pollution.

However, the West Zone, is where they locate the best green spaces of the ZMG, in which tourist corridors are generated with an adequate public mobility, with an accessible communication to the historic center of Guadalajara.

In relation to the water resource, a reflection is detected in the tourism sector in Lake Chapala, a natural heritage that attracts thousands of visitors from Guadalajara, other national and international entities. Despite the rains from 2013 to 2016, the lake has not recovered its maximum level at 7.80 meters, in 2012 it was 4.88 meters. in 2013 it was 4.02 meters. and in 2016 it was 4.21 meters. Data provided by the State Water Commission, Jalisco (2016).

The problems that Lake Chapala has are not completely ignored, there are associations that defend the lake against the aggressions it receives from the industries, hence the importance of linking the fivefold helix.

In relation to soil, evidence of serious soil contamination problems has accumulated in areas surrounding the ZMG, especially in the case of the Tesistán Valley in the municipality of Zapopan, which entails the risk of contamination of aquifers, due to leaks in containers of hazardous materials, as well as by continuous spills of lubricants, organic solvents and other types of substances, due to improper handling practices, mainly gasoline service stations, auto transport workshops, bus stations and terminals and various industries.

Solid waste is generated every day as a result of human activities. In the past, most of the waste was in a greater proportion of organic type and the volumes that were generated could still be absorbed by the natural processes of the earth. Nowadays and since the moment of the industrial impulse, the quantities of garbage increased as a result of urbanization processes and changes in factory production methods. It should be noted that the ZMG generates 5,000 tons of garbage a day and unfortunately it is not recycled.

Concerning visual contamination in the ZMG, it is affirmed that there are streets in Guadalajara that have a high number of advertisements. Another example is when you leave the airport to Guadalajara spaces where there are 64 ads and on the route to the ZMG have been counted up to 150 ads.

The flora of the ZMG, is far from being described completely despite the efforts and work of various investigations of the University of Guadalajara.

You can find large mountainous areas and extensive plains, in the multiple types of soil that contains the ZMG, and has a climate varied in temperature and precipitation that has been exacerbating the increase in temperature due to the intense urbanization in the periphery settling on land that in the past They were green spaces.

Regarding the fauna, it is pointed out that species become extinct or are threatened for various reasons, although the main cause is the destruction of the habitat. The drainage of wetlands, the conversion of scrub areas into pasture lands, the felling of forests, urbanization and suburbanization, and the construction of roads and buildings have greatly reduced the available habitats. When the fragmentation of habitats occurs in colonies of faunal species, the animal population is grouped into smaller areas. In these, the species lose contact with other populations of the same type, which limits their genetic diversity and reduces their ability to adapt to changes in the environment. These small populations are very vulnerable to extinction, and for some species these fragmented habitats are too small for a population to be viable.

It is necessary to change the tourism development model in Mexico and change it for another that is not predatory, hence the importance of the Guadalajara Strategic Plan.

With bad planning, this demographic explosion will mean that cities can not respond to the needs of their inhabitants or provide them with adequate services.

Then, environmental degradation will cause air and water to lose quality, people live crowded, health problems will arise, poverty will increase and the quality of life will deteriorate.

What to Do?

Create "sustainable cities" that accommodate urban growth, while respecting the environment. The idea is to achieve a balance between social, economic and ecological interests, and thus promote a new way of life that takes into account future generations.

How to get a sustainable city?

The process of building a sustainable city must be participatory: it is the citizen who actively rebuilds it, intervening in its planning, making decisions and applying environmental policies.

What Should This Type of Urbanization Contemplate?

From the presence of green spaces to the necessary precautions to cope with the effects of climate change. The research develops a generalized criterion of the concept of restricted urbanization of the territory to justify the changes in the urban form and functions of the growth and the worldwide extension of the metropolis. Process that has developed with a growing dynamic since 1970, with an expansion of the suburban, which has given rise to a growing polarization of new urban criteria, where the gap between rich and poor has increased as never before in history. of Jalisco and the ZMG, in addition contemporary societies cause environmental urban impacts, an energy crisis, insufficient food production, as well as overexploitation of resources, to which has been added the migration of those who lived in rural and suburban areas, to the cities.

As far as Mexico is concerned, at the dawn of the 20th century, only a tenth of the population lived in the cities; however, this proportion rose to three quarters at the beginning of the 21st century, which means a universe of 79 million urban Mexicans. Likewise, the acceleration of the urbanization processes carried out with little planning, originates negative results, among which are: The loss of identity; the generation of violence; The insecurity; climatic changes and serious consequences for the agricultural sector, since the constant and persistent migration from the countryside to the city, has been a factor identified with the deterioration of rural life.

According to the German geographer Bronger Dirk, in both developed and developing countries, a metropolis defines it as a large city with more than one million inhabitants and is spread out in a compact space, with a minimum average density of 2,000 inhabitants per square kilometer and centric mono structure. Under this criterion, the Metropolitan Zone of Guadalajara (ZMG) is located within the 340 metropolitan cities existing in the world.

The planning of urban development in Jalisco, has made multiple attempts to establish a coordinated process of urban development, as are the cases of the ZMG and in medium-sized cities such as Puerto Vallarta, Lagos de Moreno, Tepatitlán de Morelos, Ciudad Guzmán and Ocotlán, among others In spite of the fact that in most of the aforementioned localities, an urban planning regulation has been proposed, attempts of a regional territorial division in processes of 5, 10 and (12 regions in the entity, See map 1), to consider constitutional precepts, the participation of the three levels of government that have attended since 1947. It would be thought that the territorial ordering has generated a quality of life and competitiveness in the infrastructure and services of the most important cities of the country; however, Guadalajara and its area of influence (60,000 hectares, and more than 5 million inhabitants) Metropolitan Region of Guadalajara (RMG), which also includes the municipality mentioned above, those of Zapopan, Tlaquepaque, Tonalá, Tlajomulco de Zúñiga, El Salto, Juanacatlán, Tala and Zapotlanejo.

It is necessary to rethink the pertinence of these traditional instruments, if we intend to carry out strategic actions with a long-term perspective, which offer effective solutions to the future challenges of the ZMG and its surrounding region.

It is a true reality that the ZMG and its immediate surroundings during the last 22 years have lacked an authority capable of elaborating and executing urban development plans and programs with an orderly, coordinated and effective approach. It should be noted that from March 1, 1995 to December 2017, a lack of coordination of the federal government prevails with the municipalities that plan and execute isolated and unrelated policies, even though the governments are from the same political party. It is imperative in this situation, to constitute an authority with the necessary characteristics to organize and coordinate projects and policies of regional scope in a long term, but without neglecting citizen participation.

Likewise, it is detected that the ZMG does not have an updated legal norm that orders the urban development process and defines the hierarchy of competencies of the different instances of federal government: Ministry of Social Development (SEDESOL), National Water Commission (CONAGUA)), Secretariat of the Agrarian Reform (SRA), Ministry of Finance and Public Credit (SHCP).

Of the state government: Secretariat of Urban Development (SEDEUR), and the municipal governments. There is an abundance of laws, regulations and ordinances that are partial and aim to define norms and urban models. Sometimes they are opposed to each other and do not produce the concurrence of programs and actions for the achievement of development objectives and goals. Faced with this situation, it is necessary to enact an up-to-date legal framework and a regulatory improvement, in line with the reality we live in and that can guide planning with research that involves a contemporary city, without omitting its socio-economic, cultural and political complexity.

It is paradoxical to find in the case of the ZMG, that the different planning instruments of federal and local order that govern urban development, apparently keep congruence with each other, but the reality is that they have not led to a coherent distribution and growth of the structure urban, infrastructure networks, transportation systems and roads that were mostly conceived 70 years ago, evidently with a not very large vehicular traffic, compared with the current vehicle template of more than 1 million motor vehicles that circulate for the metropolis. In this situation it is observed that there is not an equitable and sufficient distribution of the urban equipment as shown in the vertebral axis of the Independence Avenue from the south zone to the north zone and vice versa, as well as the traffic inflow of Javier Mina - Juarez – Vallarta, Avenues from the East to West Zone.

The concern must be focused on the existence of a truly integrated orderly development process of the ZMG and not on the congruence of the planning instruments that are ultimately not respected or implemented in practice.

In view of this situation, it is necessary to prepare a long-term strategic plan, with the participation of all public and social actors that participate in the development not only of the Metropolitan Area, also the regional area where the municipalities will be involved. The ZMG has been living for 20 years with insecurity due to failures in the policies and institutions of law enforcement, which have altered the urban cohabitation and the implementation of social movements that force the use of public force.

The most appropriate way to increase the level of life, income and well-being of the inhabitants of the ZMG in an agile way is to adopt different vocations as an intelligent region, but with new attitudes that allow to attract with effectiveness the new investments and activities that revitalize the growth and the development process of the region.

The inhabitants of Guadalajara are fortunate to have a cultural heritage and an invaluable historical heritage that must recover and coexist harmoniously with contemporary urban development. However, the increase in rescue programs in the historic center of Guadalajara, which began in 1995, has not been undertaken comprehensive projects and urban image not only from abroad, but inside the farms, which benefit the 420 blocks registered in the Historic Center and only 6% are favored.

In this sense, citizen participation is important for the actions of those responsible for urban development policies to be fulfilled: The processes in urban participation are neutral actions in their forecasting and orientation phase, which results in planning, programming, budget and evaluation in the works and in the activity of the government areas.

The urban planning of Jalisco and its macro and micro environment is a sine qua non condition that must establish the criteria that must be carried out, be carried out in an orderly manner, and that the economic resources (even if scarce) should be applied as planned. Under the terms of the law, the planning of the urban development of population centers is a concurrent task of the states, the municipalities, and this is where citizen participation must be included in a fundamental way, in accordance with the competencies that determines the Political Constitution of Mexico.

Proposal of the Model Methodology of Quintuple Helix

According to the model of the fivefold helix that intensifies the activity of the academic line as it has been the reforestation of arid zones in the protected areas of the rights of the university students, economic sciences Administration and university center of art architecture and design that have been reinforced around 250 hectares from 2010 to 2016.

The relationship between the local authorities has been implemented reforestation actions and qualification of service and minimum equipment of natural protected networks by the secretariat of environment and territorial development and by the municipalities of ecology of the municipality of Zapopan.

In the third part. Information. On the bottom. The information has been incorporated into the company. universities and federal and state government to reforest the aerial networks as the case of the spring forest and the sentinel forest, these four words have become the fifth representation represented in the natural environment in balance.

CONCLUSION

It is known that in the Guadalajara Metropolitan Area the state of the green areas, presents serious shortcomings both in established areas and distribution. In the other hand it is known that we do not reach the WHO recommended average (9 M^2/inhabitant) neither the UN (15 M^2/inhabitant) nor the German standard (17 M^2/ inhabitant;). ONU/HABITAT (2016).

One of the proposals concerning to the WHO recommendations, German Agency and UN, is that researches could be carried out on the reduction of green areas surfaces and conduct management studies of existing arboreal species, in order to reform regulations public politicians implementation. Yves, R(2015) Of the municipalities about the management of these areas, and fulfill the pruning and removal regulation issued by the SEMADES (Secretary of the environment and territorial development) of the government of the state of Jalisco.

The article has mentioned multiple factors that affect the loss of biodiversity, air pollution and water, climate change, risk of disease, food security and deforestation, among others. These challenges in the Guadalajara region are leading current administrations to of December 5, 2018 to rethink work schemes to focus on a region that covers an area of influence of 60,000 hectares with an application of the bioeconomy as a strategy of social and economic growth oriented towards sustainability and the efficient use of biological resources that it is feasible to enhance research activities by the new administration of the universities and in the specific case of the University of Guadalajara as of April 1, 2019 where there will be changes in the general rectory and its thematic centers, which generate new approaches with which the bioeconomy copper more relevance.

Within the activities of bioeonomy associated with green areas will contribute to various approaches that highlight the reforestation of 2, 500 areas in the Guadalajara region that will not only benefit the health of its five million inhabitants but will complement it with a sustainable production of healthy foods, a more dynamic development of bio-energies, the generation of new agricultural bio-products, generation of new bio-industries with a sustainable use of biodiversity in the Guadalajara region. According to criteria of the University of Guadalajara, the bioeconomy is an emerging technology that is supported by the organization for cooperation and economic development (OECD) with the timely application of the application of science and technology by public universities and private of the

Guadalajara region. At the metropolitan level of Guadalajara at the beginning of the 21st century, a growing industrial application of the most modern biotechnology begins. Despite the success of small and medium enterprises (PiMEs) are those that have dominated the structure of biotechnology.

The bioeconomy has multiple applications such as reforestation, the generation of new management and innovation dynamics with the order of the Guadalajara region that will allow a greater application to the agricultural sector that together will contribute to reduce climate change and food security due to the growth of the population that will allow higher production rates that generate more productive and sustainable resources in a sector to which it is dedicated as a means of support, in addition with the bioeconomy will have a better health with the innovation of the bioeconomy as well as the generation of developments that contribute to recover the soils in an approximate area of 30,000 hectares and increase the quality of the water by 50 million cubic meters.

In the region of Guadalajara with the presence of the mouth of spring and the Oblate canal will contribute to the bioeconomy to have less risk of pandemics reduce climate change a greater diversification of food security, energy security and environmental sustainability that lead to the region of Guadalajara to rethink its work plans to focus on a more sustainable country, articulating joint programs and policies with the new administration of the current president of the country Andrés Manuel López Obrador, by the universities and the specific case of the University of Guadalajara in its curricula will adopt the bioeconomy as strategies of bioeconomic growth oriented towards sustainability and the efficient use of biological resources that enhance research activities linked to public sectors, entrepreneurs and society in general that together generate new bioproducts, new bio inputs agricultural, bio drugs, bioplastics, generation of new bioindustries and biorefineries that present an opportunity to add value to traditional sectors that are still identified in the region of Guadalajara.

With the participation of a landscape architecture and tourist focus can positively impact the ecological line agricultural areas, food, textiles, chemicals, construction with Ecotechnics, better health, cleaner energy, which has a potential for enter the dynamics of the bioeconomy and the application of biotechnology that contributes to developing innovative processes, products and services, taking advantage of the benefits of the natural resources of the Guadalajara region that generate opportunities for new business and economic progress.

One of the restrictions of the bioeconomy with respect to the Guadalajara region is that there are not enough diagnostic tests or bio devices that show with greater objectivity the loss of biodiversity, air and water pollution, climate change and risks to pandemics, but there are strengths as with the participation of the government, businessmen, universities and the social sector considering the environmental quality of the Guadalajara region, it will be feasible to apply developments with new alternatives of cleaner and more sustainable energies with the use of associated bioactive ingredients to health and well-being, specialized nano-biotechnology for new processes, and biotechnology in an interdisciplinary way, seeks to reduce dependence on non-renewable natural resources for the 2019-2024 cycle.

Taking as reference the regional context of Guadalajara and the situation of its green areas, it can be affirmed that there is a potential to enter dynamics of the bioeconomy and the application of a biotechnology specific to the place that facilitates the development of processes, products and services innovators taking advantage of the benefits of biological resources, genetic, residual biomass, the incorporation of knowledge by universities and innovation by employers with the due consent of the government at its three levels as federal state and municipal, which jointly generate new business opportunities and real economic progress.

REFERENCES

Anaya, C. M. (2008). *Análisis espacial de los espacios abiertos recreativos, de uso público en el área metropolitana de Guadalajara*. Jalisco, México: Universidad de Guadalajara.

Bautista, E. G. (2015). *La importancia der la Vinculación Universidades-Empresa-Gobierno en México. Volumen 5, Número 10, Revista Iberoamericana para la Investigación y el Desarrollo Económico*. RIDE.

Reglamento de Parques, Jardines y Recursos Forestales para el Municipio de Guadalajara. (2014). Gaceta Municipal de Guadalajara.

Etzkowitz, H., & Leydes, D. (2000). *The Triple Helix. University-Industry-Governement. Relations; A Laboratory for Knowledge*. Based Economic Development Easst-Review.

Hábitat & Keever. (2005). *Ciudades "Sobre indicadores de Sustentabilidad Urbana y Observatorios*. Red Nacional de Investigación Urbana.

Marco, O., & Edmond, P. (2016). *La ségrégation urbaine. Editions. La Découverte*. Paris: Francia.

Miranda, C. (1997). Filosofía y Medio Ambiente. Una Aproximación teórica. Taller Abierto, México, D.F.

Muñoz & Carmen. (1996). Principales tendencias y Modelos de la +Educación Ambiental en el sistema escolar. Revista Iberoamericana de Educación, 11.

Raibaud, Y. (2015). *La ville faite par et pour les hommes*. Editions Belin, Paris, Francia. Retrieved from http://iitej.blogspot.mx/2012/11/cantidad-de-areas-verdes-de-guadalajara.html#more

Thierry, P. (2016). *Terre urbaine. Cinq défis pour le devenir urbain de la planéte. Editions. La Découverte*. Paris: Francia.

KEY TERMS AND DEFINITIONS

Bioeconomic Valuation: Set of economic activities that obtain products and services, generating economic value, using biological resources as a fundamental element with a sustainable approach.

Cities: The main activity is industry, commerce, and agricultural exploitation.

Insecurity: Absence of security that an individual or a social group perceives regarding their image, their physical and/or mental integrity and in their relationship with the world.

Quintuple Helix: Model that supports economic and financial aspects of the environmental challenges inherent in the causes and effects of climate change that require political and social measures that are implemented in coordination with the government, businessmen, universities, and society in favor of the environment.

Society Exclusion: It refers to the processes and situations that impede the satisfaction of the basic needs of people (work, housing, education, access to healthcare) and their participation in society.

Stress: State of mental fatigue caused by the demand for a much higher than normal performance; it usually causes various physical and mental disorders.

Urban Green Areas: They are urban spaces or of the periphery predominantly occupied with trees, bushes or plants, that can have different uses, either to fulfill functions of recreation, recreation, ecological, ornamentation, protection, recovery, and rehabilitation of the environment.

Urbanism: Study of the planning and management of cities and territory as well as the planning or sustainable design of a population.

Chapter 8
Application of UAV Technology to Planning Study on Chinese Villages in Guanzhong

Yuan Zhong Cai
Xi'an University of Architecture and Technology, China

Feng Wu
Xi'an University of Architecture and Technology, China

Jing Li
Xi'an Polytechnic University, China

Jin Wang
Xi'an University of Architecture and Technology, China

Mei Huang
Xi'an University of Architecture and Technology, China

ABSTRACT

Driven by the state strategy of rural revitalization, Chinese rural areas receive unprecedented opportunities for development. However, China's Guanzhong region faces numerous problems in its rural planning research, such as 1) lack of terrain maps of most villages, 2) satellite maps collected from open platforms are inaccurate and fail to support a more detailed spatial analysis, 3) data and information are 2-dimensional, 4) data collection is inefficient. And, most villages consist of several village groups that are usually 400~500 m apart. Areas of Guanzhong are located on the plain, with low architectural height and an excellent environment of net clearance. In addition, there are no large-scale factors, mineral areas, and industrial facilities, which means low interference from the magnetic field. Compared with urban regions, such rural areas have a better work environment for UAV and better conditions of collecting needed data.

DOI: 10.4018/978-1-5225-7958-8.ch008

INTRODUCTION

In recently years, the development of Chinese rural area have been focused by the national: *Urban and Rural Planning Law of the People's Republic of China* have been promulgated at the year of 2008, which is confirm the rural status equally to urban by the law; The model of Construction of Beautiful Village have been promoted strongly by Chinese government; as well as *The Report of 19th National Congress of the Communist Party of China on October 24, 2017* that is declare the balance urban and rural development, rural development has been raised to a new height. All the changes of policy mean the status of rural development have been changed deeply, the rural area would be get more and more chances form national police.

Guanzhong area is located in central Shaanxi Province and the geographical center of inland China. The distance stretching from the east to the west is about 300 km. Guangzhong area reaches Tongguan in the east, Baoji in the west, Qingling Mountains in the south and North Mountain ranges in the north. Located between Hanguguan and Dasanguan, Guanzhong gained its name in ancient China (meaning the central pass in Chinese). Weihe River is known as Qinchuan. Hence Guanzhong area is also referred to as "800-li Qinchuan". This area which has been easy to hold but hard to attack since surrounded by mountains that is good narural military barrier on three sides. The Guanzhong area has fertile and abundant land, with well-developed river system and flourishing agricultural, so, it had been one of the most rich agriculture areas in China from ancient times to the present, as well as good traffic condition, it is also a densely populated area in China. There were 13 dynasties have established capitals in the Guanzhong area which have the longest capital history of 1,200 years in Chinese history. since the Zhou Dynasty (1136 BC) because well military defense conditions and agricultural foundation. So Guanzhong area that was the starting point of the Silk Road with profound historical and cultural significance, witness one of the most prosperous periods in Chinese history. So, Guanzhong area become the key area where is to be study the development mode of Chinese cities and towns.

Under the rapid urbanization development in China and the strategy of national rural revitalization, as the center inland area of China, the Guanzhong area have been became the has been taken the steps to speed up large-scale village planning where in poor regions zone in a short time and in a wide range. Nevertheless, it is hard to finish the large-scale village investigation with common pattern that is to acquire and analysis date because the investment and the local economic, consequently, it would be provided the new ideas that adapted to the new era by the new technical route and analysis procedure.

Geographical Characteristics of Guanzhong Area

Guanzhong area is located in the south of Shanxi and Shaanxi basins. It is a basin that including the Weihe Plain at the northern foot of Qinling Mountains and the Weihe River Valley in Shaanxi Province along the Weihe River. The northern part of Guanzhong area is the Loess Plateau of northern Shaanxi, as well as the southern part is the Qinling-Daba Mountains. The area near the Qinling Mountains forms stepped steps. The central part of Guanzhong area is close to the Weihe River region with flat terrain and dense population. The village construction space presents scattered distribution in the vast farmland because the plain terrain in the central part, convenient traffic links and small terrain restrictions (see Figure 1).

Topographical Characteristics of Villages in Guanzhong Area

Most villages in Guanzhong area are located on the plain, whose construction zones are free from the impact of terrain and landform. In addition, villagers which form

Figure 1. Villages in map of administrative regions of Guanzhong
(Data source: drawn by the author)

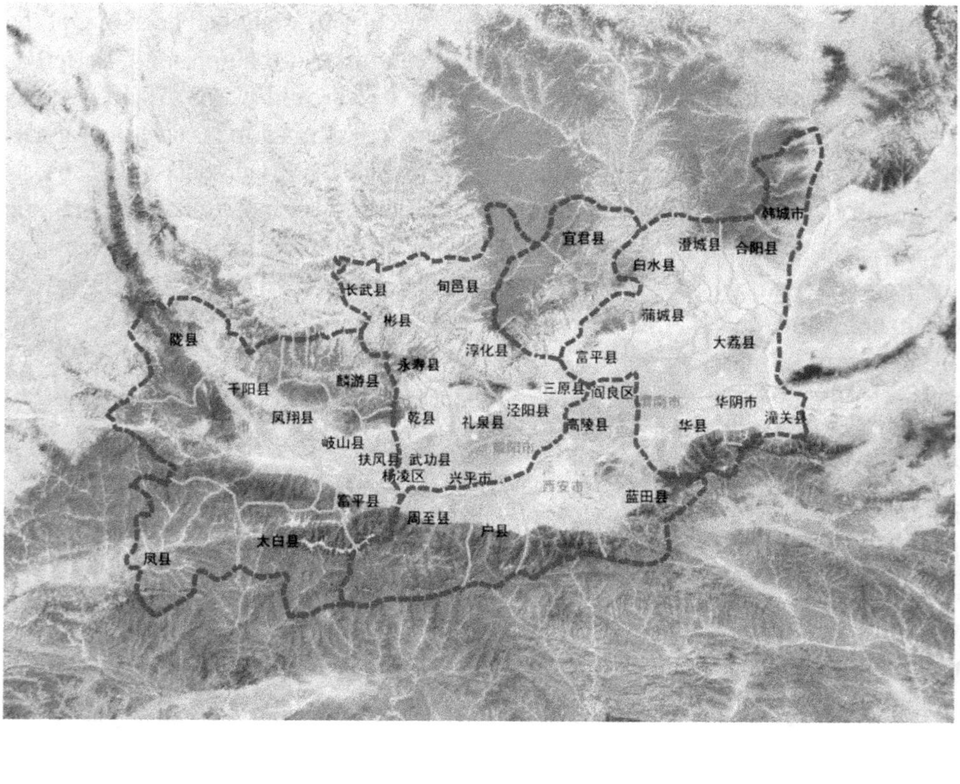

is block mass layout along the line of transportation. Then, owing to long human timescale in Guanzhong area, the most of the Guanzhong area land except for rivers and a few land not suitable for farming are farmland that around the village resideyntial areas. Additional, the rural productive forces depend on manpower and animal power because traditional farming mode in the past, the farmer walk form residential area to farmland for work, so the number of cultivating radius that is the distance from residential to farmland between 400m and 1100m, which is not far away. One village consisted by several residential areas, the distance between each residential area is 400 m ~ 500 m, and the village residential area morphology presents uniform distribution. (see Figure 2 and Figure 3).

Figure 2. Satellite map of Wu Village, Fuping County, Guanzhong
(Data source: drawn by the author)

Figure 3. The plane of land use in Wu Village, Fuping County, Guanzhong
(Data source: drawn by the author)

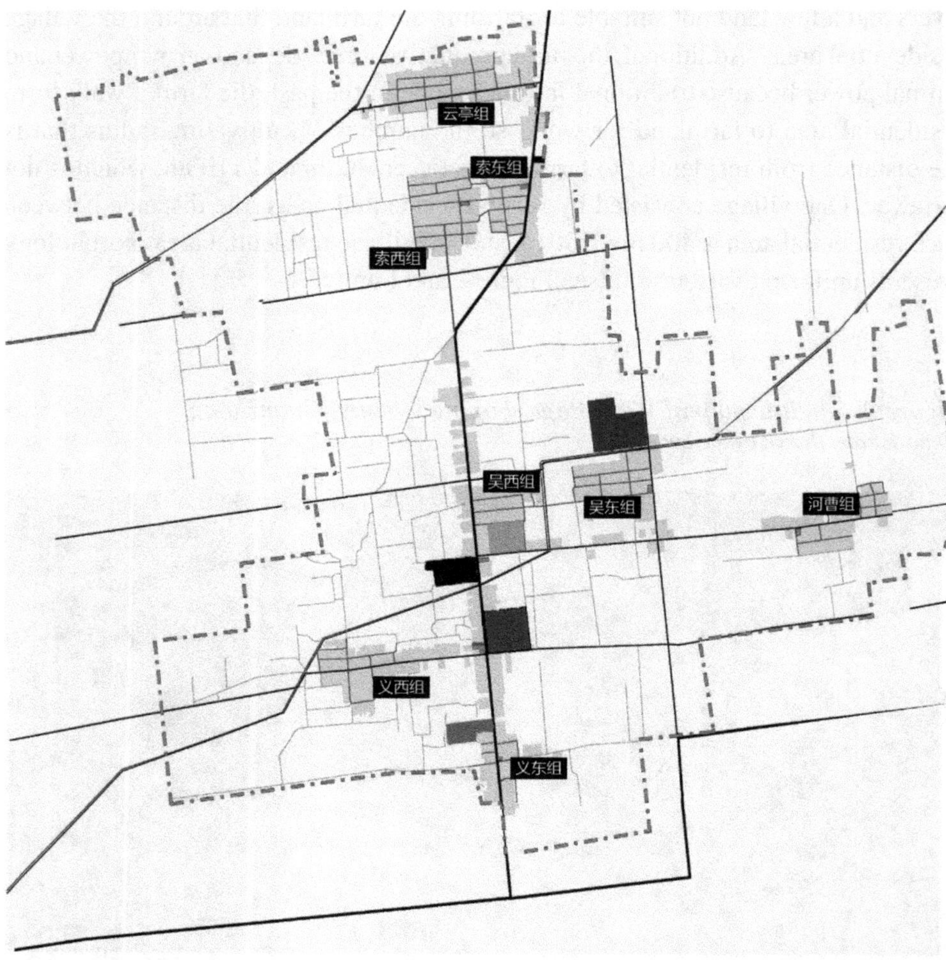

Problems in Collecting Geographical Data for Village Planning of Guanzhong

The problems currently faced by the research in the front of rural planning in Guanzhong region are listed as follows:

1. There lacks terrain maps of most villages. Traditional rural terrain mapping has three deficiencies: (1) it would be lengthen the time of planning research

because ground survey terrain map ; (2)The terrain mapping fail to cover all of area of village, such as all production of farmland, and ecological spatial, and only the residential area is covered by terrain mapping. due to the restrictions of economic conditions; (3) The terrain mapping have low precision so that lack partial space elements, including the characteristic trees and stones in villages.

2. The satellite maps which lack of detailed spatial information collected from open platforms are inaccurate, fail to support a more detailed spatial analysis and reflect the latest changes in the aspect of construction of villages
3. Data and information are 2-dimensional. The current research generally integrates all elements and information based on 2-dimensional terrain maps. Hence all elements of villages are represented on 2-dimensional drawings, making it inconvenient to study stereoscopic spatial relations.
4. Data collection is inefficient. Restricted by current conditions, traditional survey modes collect all spatial elements and information from on-site drawing marks, photography and recordings, which would be consume much of manpower and time for on-the-spot surveys. Hence current collection of initial information for village planning study is inefficient as well as the mode of spatial analysis is backward.

Accordingly, Not only the ineffective of collecting preliminary information on village planning research, but also the methods of spatial analysis were lagging behind the times, therefore, it difficult to launch the research work of village planning rapidly in a short period of time.

BACKGROUND OF UAV TECHNOLOGY

Introduction to UAV

UAV is the abbreviation for Unmanned Aerial Vehicle. Due to a light weight, a small size, low difficulty for technological platform and low flight costs, UAV technology has enjoyed rapid development in recent years. UAV generally consists of the flying control system, dynamic system, data transmission system and ground workstation system. UAV either flies according to the flight signals sent by the flying control system or following the flight route set in advance. The UAV that camera is installed camera could shoot photos according to needed angles along the flight route and transmit such photos to the ground workstation via the data transmission system. In recent years, with the development of UAV technology, the cost of aerial flight control equipment has been gradually reduced, which makes it possible to use this technology in the field of rural planning.

Principle of Oblique Photography Modeling

The information of position and appearance of the ground objects could be reflected by Oblique photography technology that has three features: (1) it can obtain three-dimensional vivid reproduction data; (2) it can truly reflect the height and other characteristics enhance, such as the texture and color. So it could be making up for the shortcomings of low simulation of the traditional model. Early Oblique photography technology used manned aerial vehicles, satellites or fire balloon for photograph, although the technology is mature, it is difficult to apply to civil technology, since the cost and technical threshold is high, After the popularization of UAV technology, Oblique photography technology become more and more popular in the construction field of urban and rural since the two technologies be united in wedlock.

Oblique photography refers to an airplane-mounted or UAV-mounted camera photographs objects on the ground in five directions, including front, back, left, right and vertical (see Figure 4) and processes through the geometric correction, adjustment and multi-vision image matching of interior work to collect all-direction information and data of ground objects (see Figure 5). Simply speaking, photographed ground objects are on the same plane, while the ground objects measured with

Figure 4. Principle of oblique photography modeling
(Data source: drawn by the author)

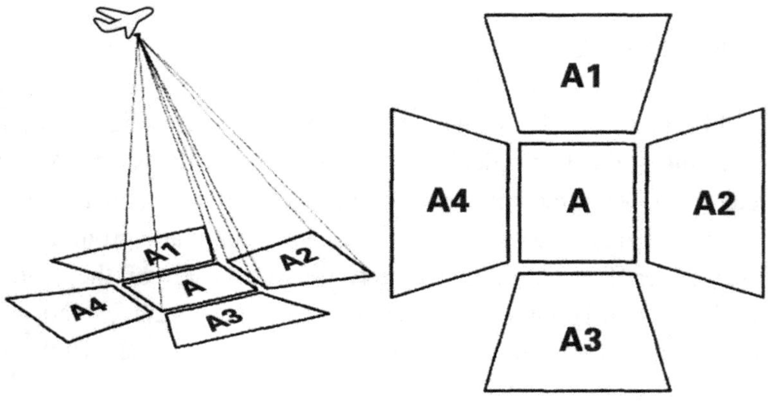

Figure 5. Computer interface of modeling software for oblique modeling
(Data source: drawn by the author)

oblique photography have true altitudes. (Zhang T.,2016) Photography with the camera mounted on an airplane led to high costs. Conversely, photography with UAV substantially reduces photography costs and lowers technological thresholds. (Yang C, 2016, p. 55)

Adaptation of UAV Technology to Rural Areas in Guanzhong Area

The rural areas of Guanzhong area are located on the plain, with low architectural height and an excellent environment of net clearance, it is suitable for controlling and flying in the larger scale, since the directly connection between ground control station and UAV without tall and complex obstacle. In addition, there are no large-scale mineral areas and industrial facilities, which means low interference from the magnetic field. Compared with urban regions, such rural areas have a better work environment for UAV and better conditions of collecting needed data.

Technical Procedures of UAV Technology in Rural Planning

Work Procedures

All work procedures are divided into five phases, including Survey Plan Formulation, On-site UAV Data Collection, Data Analysis & Arrangement, Model Output and Data Analysis.

The first phase is Survey Plan Formulation, which aims to clarify the utilization type of needed data and data accuracy, such as whether to adopt orthographic projection plane or geographical information with elevation, to determine the flight routes and height in a flight plan. In this step, the flight routes should be translated to flight coordinates, which ensure the flight task feasibility.

On-site UAV Data Collection refers to UAV which arrives at the survey site flies in accordance with the planned flight route after observing on-site flight conditions (see Figure 6). It is necessary for UAV to adjust its flight posture in flight according to the conditions on the site. The open space should be selected for launch site,

Figure 6. On-site data collection
(Data source: drawn by the author)

neither tall obstacle that reduce the control signal of UAV, nor the magnetic field that interference the GPS signal, afterwards, the UAV started take off condition after the have finished it self-checking, it would be enter aerial photography situation for picture which file contain location coordinates and camera angles in the pre-established routes.

The stage of data analysis is also go by the name of the data processing stage, the precise exterior orientation elements would be ensured by Oblique Photography operation software operate "POS" calculation with location coordinates and camera angles, and the next step is the aerial triangulation. The core theory of software operation is that calculating the positions of the same points on different photos in relation to the photos at different positions to judge the coordinates of such characteristic points in the real world. After the system determines the coordinates of spatial points, it will form a point-cloud model that Digital Surface Model (DSM) data of super high density point cloud based on real image, then, using the third-party model editing software to repair the defects of the model, improve the refinement of the model, finally, model would be returned to the model structure, (Wang,Y., 2008) finally, the computer system then makes form fitting of point-cloud mode to construct a digitalized terrain mode (see Figure 7).

The types of data input refer to converting photo files into Digital Elevation Mode (DEM)data, Digital Surface Model (DSM) data, DOM data or MESH mode data.

Data Utilization Mode

All model data is input into different platforms for different analysis aims. The current objectives of data utilization include geographical data abstraction, current spatial analysis and simulation of future rural planning.

- *Abstract Geographical Data*

The data arrangement through digital elevation mode converts DEM file into a DWG file. It then converts a large number of files collecting geographical information

Figure 7. Framework of model data analysis mode
(Data source: drawn by the author)

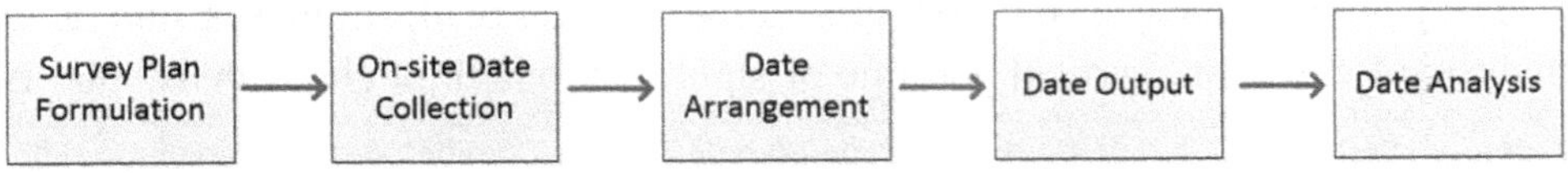

Figure 8. Orthographic projection of Donggongcheng Village in Guandong
(Data source: drawn by the author)

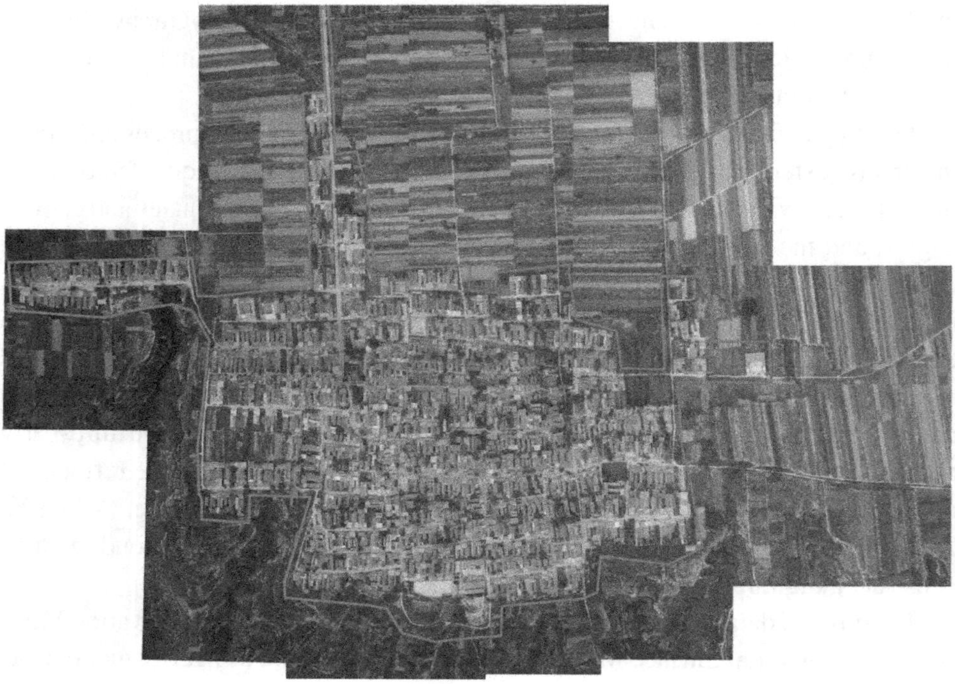

on the artificial ground site into aerial photography to get aerial photos within a short time and generate terrain files on indoor work. Precisely geographical dater would be got if serval control point which coordinates have determined the overall location relationship of a region should be placed in the research area in preliminary data collection phase.

- *Status Spatial Analysis*

The large amount of spatial information in the real environment can be centralized into the one model by collecting more massive data of spatial location information in the field. In this way, it transform the traditional spatial analysis that remodeling with terrain maps without more detailed information of ground surface into analysis the accurate model directly.

The simulated analysis of geographical information platform makes simulated analysis of elevation, vertical and line of sight relations through "ArcGis" that is

Figure 9. Terrain map formulated by digital elevation model
(Data source: drawn by the author)

Figure 10. Framework of analysis mode for model data
(Data source: drawn by the author)

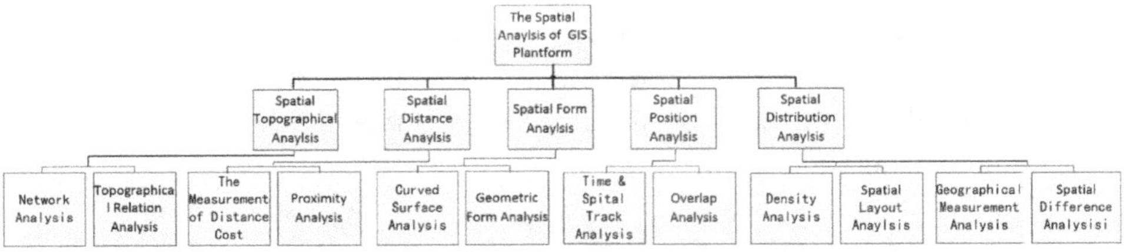

geographic Information Analysis platform. The evaluation researches and judges the regions suitable for future village development with current land conditions, and it is provide a basis for future village development.

- *Future Plan Simulation*

The future planning design can be pre-judged by aerial survey model. The new models can be incorporated into real survey model through simulation platform to compare the present situation and completion vision by the intuitive manner.

Difficulties in Application of UAV Technology

It is still in the exploration stages which have some problems by using UAV technology at rural planning survey of Guanzhong area, so, there are some difficult points in applicate the UAV technology.

1. Environmental Sensitivity

UAV has high requirements on flight. It highly sensitive to weather factors, including wind, rain, snow, thunder and so on special weather condition, which are adverse conditions for normal weather taking off or flight. Secondly, abnormal magnetic field or electromagnetic field disturbs GPS system, which also makes it disable for UAV to safe fly.

` *2. Improvement in Entry System*

The current UAV flight requires the controller to abide by relevant laws and regulations. Before operating the UAV, the controller with flight license should report route and time to management departments for approval. Hence this technology has high thresholds on professionalism and entry.

3. Difficulty for Data Analysis and Utilization

The current design mode for rural planning analyzes and surveys with traditional satellite maps and terrain maps. It is accustomed to 2-dimensional survey and analysis and hasn't developed the habit of working with 3-D model. Moreover, most UAV application remains superficial.

4. Difficult For Edit Points Cloud Model

The points cloud model formed by a large number of key points which form the basic characteristic of the measuring object, it has the features of large amount of data and Irregular shape that is linked by each point to form a complex topographic surface. Big date would impact the speed of analysis. The second, all shape elements in the whole model, in other words, all elements are on one layer, it is so hard to distinguish different types of elements, such as buildings, roads, vegetation and so on, it needs to be implemented in the software later. And the workload is heavy. So, the both characteristics of points cloud model, which makes it difficult to edit the point cloud model later.

CONCLUSION

Unmanned aerial vehicle (UAV) is a very important tool to acquire ground spatial information automatically, intelligently and specially. Although, some technical difficulties have not been overcame in the field of rural planning, the defects cannot belittle virtues, its potential has not been fully exploited, therefore, the schema of preliminary research of rural planning would be promoted by its advantages that is maneuverability, speed and economy in future.

In the future, the date research methods should be targeted and focused on following aspects: 1 The different depth levels rural space model should be established by the points cloud model that is collected with the help of UAV Oblique Photography Modeling, which is decided by re-pectinate the key problem of rural planning and design; 2 The special sensor could be mounted in the UAV, so as to obtain. targeted data that is key date for rural planning, such as different crop distribution data sources could be collected by multi-spectral cameras; or the date of village microclimate environment could be obtained by thermal imaging cameras; 3 Image recognition research should be developed, so that different types of elements data would be distinguished by analyzed by system automatically. Such as building, roads, water, farmland and so on could be automated distinction. Based on the above future research, the UAV technology would be worked at rural planning with high efficiency at high altitude, and it will provide a more scientific and rational basis for judgment based on the characteristics of rural planning and design.

Basin on the high effective technology of UAV and Oblique Photography, the information of ground spatial would be more cheaper than before, traditional model as well as the analysis pattern would be promoted, further, the rural planning of Guanzhong area would be more and more efficiency in the earlier stage for date collected and analysis, the researcher could easily get the very comprehensive spatial information of village without long time investigate at the field investigation, so it has well application prospects in the field of rural planning at the Guanzhong area as an important tool. At the same time, the time that investigation and analysis at the aspect of spatial would be less and less, so the more and more time could be used at survey the villager who is the key element in the rural planning, so it has play a positive role for improving the quality of rural planning.

ACKNOWLEDGMENT

This research was supported by the Foundation of Shaanxi Educational Committee [grant number 17JK0460], and the Natural Science Basic Research Plan in Shaanxi Province of China Program [grant number 2016JM5084], and the National Natural Science Youth Foundation of China[grant number 51608417].

REFERENCES

Gu, J. (2016). Application of UAV Photography to Ownership Affirmation, Registration and Certification for Right to Manage Contracted Rural Land—Take Fengyang County as a Case. *Bulletin of Surveying and Mapping,* (5), 94-99.

Wang, Y., Schultz, S., & Giuf, I. F. (2008). Pictometry's proprietary airborne digital imaging system and its application in 3D city modelling. *International Archives of Photogrammetry, Sensing and Spatial Information Sciences,* (37), 1065-1069.

Zhang, T. (2014). Application of UAV Photography to Urban Planning Construction. *Urban Geotechnical Investigation & Surveying,* (2), 99-101.

KEY TERMS AND DEFINITIONS

Guanzhong: One of the most important Chinese cultural birthplaces corresponding to the lower valley of the Wei River.

Construction of Beautiful Village: One of the most important Chinese policies is based on "Beautiful China" policy. Since the 18th National Congress of the Communist Party of China, the target is development of industry and the improvement of living environment and rural culture base on the background of construction of ecological civilization.

DWG File: One of file formats used for storing two- and three-dimensional design data and metadata. It is the native format for several CAD packages including DraftSight, AutoCAD, IntelliCAD, and so on.

Chapter 9
Developing a Sustainable Eco-City in Pre-Olympic Tokyo:
Potential of New Methods and Their Limits in an Urban Era

Mariko Ikeda
University of Tsukuba, Japan

ABSTRACT

This chapter intends to elucidate the emergence of sustainable urban development in Tokyo in light of the upcoming 2020 Olympics by exploring various administrative and commercial practices, such as urban development plan with rooftop and wall greening or river projects in the inner city. The research methods involved a review of the empirical literature, an analysis of existing statistical data, and a detailed examination of case-specific data collected in a field survey. This chapter concludes that since Tokyo Metropolitan Government encourages urban greening projects as a solution of urban heat island from 2000s, utilization of "green spaces" in the landscape design of commercial and office facilities is gaining attention. This chapter concludes that various practices for sustainable urban development in Tokyo, which faces a restructuring process in light of the 2020 Summer Olympics, exist and that some of these could be further developed by the private sector.

DOI: 10.4018/978-1-5225-7958-8.ch009

INTRODUCTION

Environmental Problems in the Urban Era: Overinvestment, Overpopulation, Overshoot

The concepts of "city" and "nature" are perceived as intrinsically contradictory, and authors who stand for the latter have criticized cities for their negligence of and dominance over the environment. Moreover, on a global scale, the concentration of populations in urban areas is growing. It is estimated that in 2030 more than 60% of the world's population will live in cities, and 27% will reside in cities with at least 1 million inhabitants (UN Habitat, 2016). The agglomeration of gross domestic product (GDP) in cities escalates accordingly and 80% of the world's total GDP is concentrated in urban areas. Paris, for instance, houses only 16% of the French population but produces about 27% of France's GDP (UN Habitat, 2016, p. 27). This condition is known as the 'overdevelopment, overpopulation, overshoot' issue, suggesting the excessive burden to the natural environment caused by urbanization (Butler, 2015). The United Nations has therefore adopted a new 2030 Agenda for Sustainable Development, which presents 17 goals that replace the previous Millennium Development Goals (United Nations, 2016, pp. 38-39). This agenda includes several new elements, such as making cities and human settlements inclusive, safe, resilient, and sustainable. Furthermore, it demands the development of low-carbon societies and green technology or the replacement of conventional energy resources with renewable energy, especially in cities, which represent more than 70% of the global energy demand.

Environmental Problems in the Urban Era: Heat Island

Japan's capital Tokyo is, together with its surrounding areas, the largest megacity in the world with approximately 38.1 million inhabitants, which is more than 10 million people more than the population of Delhi, the second-largest city (United Nations, 2016). It is also the city with the largest GDP worldwide. Tokyo's population increase began during the period of Japan's rapid economic growth (1955–1973) with an influx of people from rural regions into suburban areas, which were constructed around the inner-city area; this trend of urbanization is still increasing (Figure 1). The extreme concentration of the population in the metropolitan area has led to an increase of housing, offices and commercial facilities and to a decrease of the natural environment, causing an environmental problem characteristic for large cities, the heat island effect.This phenomenon, also called 'city climate', occurs rather in the

Figure 1. Rate of net migration for the three major metropolitan areas in Japan from 1954 until 2017
Source: (e-Stat, 2017)

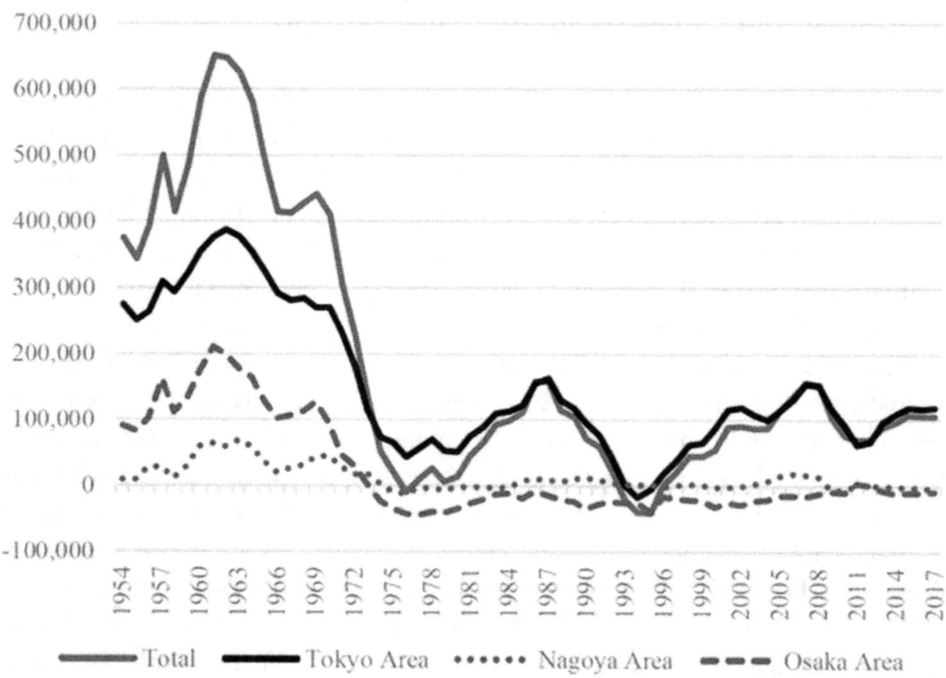

city's center than in its outskirts and is mainly caused by solar heat during the day and by waste heat from human economic activities. This artificial heat disposal is said to be equal to 10% of the sun's radiation and to have a stronger effect with increasing city size (Sakakibara, 2001). Observations of the temperature difference between the center of a city and its suburbs show that the difference during the daytime is small but tends to increase during the night because the temperature in the city center does not fall (Maeda, Kondo, Akikawa, Kaga, & Inoue, 2008; Masumoto, 2007; Mizuno, 1993). The temperature is not only dependent on such a parallel distance from a city center, but it is also said to be inversely proportional related to the altitude of the observation spot, and it is supposed that cold air in the atmosphere over the city traps the urban air in a dome.[1] For example, the urban air dome so-called temperature inversion layer above New York is situated at a height of about 310 meters (Bornstein, 1968), in the case of Japan's Nagoya it is at a height of approximately 60 meters (Aikawa, Hiraki, & Eiho, 2006). If one assumes that one floor of a building has a height of 4 meters, this would correspond to a building with 77 floors in the case of New York and to a building with 15 floors in the case of Nagoya. In other words, this phenomenon is not only related to physical factors such as anthropogenic heat but also strongly connected to social factors such as the increase in artificial building structures (Maeda et al., 2008). For example, in the city center, there exist only intensive land use such as high-rise buildings with offices and apartments, which can sustain the high land prices. This leads to the weakening of natural winds [mechanical effects], reducing humidity, generating a complex reflection of shortwave and longwave radiation between buildings, generating thermal effects that further intensify the heat island phenomenon (Kondo, 2009; Oke, 1979). In addition, this urban air dome contains air pollutants, which stem from exhaust fumes of cars and from tall smoke sources, such as thermal power stations (Uno, Wakamatsu, & Ueda, 1988). It is inferred that such an urban climate phenomenon will further change through the expansion of the suburban area with increasing population concentration in the metropolitan area (horizontal expansion) and the vertical expansion of the inner city. In other words, regarding environmental concerns in this urban era of overpopulation, case study in Tokyo, in which approximately 25% of Japan's total population are living, could show a best challenging practice.

Olympics and Paralympics: Urban Competitiveness Towards Sustainable Eco-City

In September 2013, Tokyo was selected as host city for the 2020 Olympic and Paralympic Games (from here on Olympics). During the candidacy, the bid committee for the Olympics pledged in its campaign to give priority to the environment. As reason behind this can be seen that the Organizing Committee for the Olympic Games has pushed forward actions of environmental consideration since 1994. Environmental protection was added to the Olympic Charter and the promotion of sustainable development is pointed out as mission and role (Saito, 2014). The year 2020, when the Tokyo Olympics will be held, is the target year for the reduction of the world's greenhouse gas emissions, and it is also the target year for the Aichi Targets formulated at the Convention on Biological Diversity. Therefore, parallel to the improvement of the environmental infrastructure with the latest environmental technologies the recovery of natural water and green spaces is also expected (Saito, 2014).

In addition, environmental improvement in the urban era may run into difficulties, in Tokyo in particular, where it often is difficult to reach a consensus regarding redevelopment due to its subdivided groundrights. In reorganizing the city and structure for a national event such as the Olympics, the awareness of universal standards regarding global city environments, there may be instances where certain considerations are applied for the first time to a city's environment. For example, in the reports of the Japanese Ministry of the Environment, in which all of Japan's environmental administrations are summarized, natural spaces are identified as a major factor in the competition between cities.

Accordingly, the problem of Tokyo's low green space ratio is addressed. Based on the assumption that the green space ratio of the inner cities of Stockholm, Geneva, and Berlin is 100%, London and Paris would have a ratio of 62.5%, New York would be at 50%, and Tokyo would be just over 25% (Ministry of the Environment, 2014). In this way, in the time before the Olympics, environmental problems are reconsidered together with the ongoing large-scale restructuring of inner-city areas, such as Ginza, Nihonbashi or Shibuya.

Research Purpose and Method

This chapter intends to elucidate the emergence of sustainable urban development in Tokyo in light of the upcoming 2020 Olympics by exploring various administrative and commercial practices in the inner city. The research methods involve a review of existing literature and published documents, an analysis of statistical data and qualitative interviews with authorities, and a detailed examination of case-specific data collected in a field survey. The study will first examine the historical change of environmental policies in Japan, followed by an overview of the environmental policies in Tokyo regarding their relation to the Olympics.

URBAN ENVIRONMENTAL POLICIES FOR A SUSTAINABLE ECO-CITY IN HOST CITIES OF MEGA EVENTS

Overview of Environmental Policies in Japan

Japan's environmental problems became apparent during the period of rapid economic growth that began in 1955. There were mainly seven pollution problems caused by the industry, namely air pollution, water pollution, soil pollution, ground subsidence, noise, vibration, and bad smell (effluvium). In the 1960s, the environmental damage spread throughout Japan together with the development of the heavy and chemical industries, leading to protest movements by residents against this pollution and development. In response to this situation, the Basic Law for Environmental Pollution

Control (1967), the Air Pollution Control Law (1968), the Noise Regulation Law, and other 14 laws (1970) regarding pollution (e.g. Water Pollution Control Law and Marine Pollution Prevention Law) were enacted. And in the last years of the period of rapid economic growth, the unification of the environmental administrations was planned with the start of the Environmental Agency (1971).[2] As a result of this series of laws, the pollution decreased since the second half of the 1970s. On the other hand, in a report by the OECD with the title 'Japan's experience – Did the environmental policies succeed?', the environmental policies of Japan where evaluated as inclined to the 'prevention of pollution' (Takeuchi & Lee, 1988). Therefore, the 'creation of a comfortable environment' (or environmental amenities) became the key concept during the 1980s, and in the Regional Environment Management Plan the viewpoint of environmental creation, including landscape preservation, was introduced. In addition, the relocation of production facilities to other countries due to an increased oil price, trade frictions, the strong Yen, and rising labor costs lead to a fading consciousness of pollution, and a response to environmental problems on a global scale became necessary from the mid-1980s (Lee, Harashima, Lee, & Morita, 1995). Furthermore, taking the amendment of the Urban Green Space Conservation Act (1994) as an example, each municipality became able to carry out greening measures in the 1990s.

For such environmentally friendly projects, examples from Europe, in particular from the U.K. and Germany, have always provided points of reference. For example, the establishment of communal areas, such as green spaces or public squares, since the 1970s; or the establishment of allotment gardens (Japan uses the German word 'Kleingarten') and biotopes since the 1990s. Especially the Kleingarten represents a place for city dwellers to enjoy gardening at the weekend and is modeled after Germany's green tourism. In fact, these gardens were mainly developed in the countryside under the name of promoting green tourism after the enactment of the Rural Regions Leisure Law in 1994, a time when the revitalization of rural areas was expected/desired. In addition, taking the biotope, a German invention, and its ideas about preservation, regeneration, and creation as model, public facilities at urban factory sites and schools were constructed and the conservation of agricultural areas was pushed forward. However, because the biotope is a concept opposed to agricultural productivity, it has not spread to other areas outside the rural regions promoting green tourism. On the other hand, a site-specific spread can be found in urban areas, depending on land development design and a limitation of land prices, and rather than a solution to environmental problems, such as the heat island effect and air pollution, the biotope's functions to preserve a green landscape and as a place for recreation and relaxation are emphasized.

Environmental Policies in the Tokyo Metropolis

The history of environmental policies in the Tokyo Metropolis is, depending on the point of view, old. The previously explained pollution problems were at first regional and not national and therefore the 'Regulations for the control of Tokyo's factory suburbs' were enacted in Tokyo in 1949, which were countermeasures at a very early stage even from a national perspective (Lee et al., 1995). Furthermore, a garden city plan was drawn up before the Second World War, and occasionally, town planning and an all-over environmental development occurred.[3]

On the other hand, some turning points exist when looking at the environmental policies in Tokyo after the war. One of them is the hosting of the mega event Tokyo Olympics. Mega events promote improvements of traffic and electricity infrastructure and the redevelopment in a city, and in the case of Tokyo, it greatly affects the city's environment. Therefore, I will survey the aspects of the 1964 Tokyo Olympics and the planned 2020 Tokyo Olympics from the viewpoint of the city environment.

Tokyo Olympics 1964: Developmentalism in the Postwar Era and Its Environmental Load

The population growth in Japan during the economic recovery after the Second World War and during the period of rapid economic growth modified the city environment rapidly. Beginning with the conversion of forest and agricultural land into residential land, backed by the developing industrialization through the formation of industrial complexes along the Pacific belt zone, the improvement of major roads and harbors and their connection by the construction of expressways was needed. During this time, the 55th International Olympic Committee general meeting was held in 1959, where Tokyo was selected as host city for the 1964 Olympics. Postwar Tokyo with its revitalizing economy devised an urban planning for the hosting of this large event five years later and underwent rapid changes. The urban planning of this period is summarized according to Echizawa (2014) in Table 1.

The characteristic of this urban planning was that land that was requisitioned after the war by the American military was returned to Tokyo metropolis under the nominal cause of the holding of a national event with high publicity called Olympics. For example, Yoyogi in Shibuya Ward where the Olympic village was developed, was used until World War II as Yoyogi parade ground and was requisitioned after the war by the United States for use as residential area "Washington Heights" for Americans and used until the early 1960s. The Japanese cabinet decided that the same grounds should be turned into a forest park after the end of the Olympic Games,

Table 1. Developmentalism for Tokyo Olympics in 1964

Development	Urban greening	Infrastructure
Re-use of Olympic village as an urban park	✓	
Construction of Olympic village		✓
Deregulation of City Planning Act (parks and greens) and construction of international accommodation		✓
Roadway and highway infrastructure		✓
Development of rail infrastructure		✓
Construction of the National Stadium		✓

Source: (Echizawa, 2014)

therefore enabling the return of the land and the securing of a forest park which is in use as forested green space until today (Ozaki, 2002). Furthermore, based on the Capital Region Improvement Law in 1956 the First Metropolitan Area Improvement Plan (1958) followed a pre-war plan to establish a green belt of 5 to 10 kilometers with an area of 108,000 hectares around the city area of Tokyo (23 wards), which was intended to supply greenery, secure land for public use, and preserve superior agricultural land for the provision of fresh food (Ministry of Environment, 2014; Ministry of Land, Infrastructure, Transport and Tourism, 2011). This plan was completely disapproved of in the Second Metropolitan Area Improvement Plan and eventually not realized. However, the establishment of a green belt between the core city and the satellite cities had the aim to reduce the urban sprawl and had used the plans for Greater London as reference (Takeuchi & Ishikawa, 2008; Maruyama & Nakagawa, 2004). In this way, the Olympics were taken as opportunity to plan the conversion of large tracts of military-used land that dotted Tokyo into green spaces of which parts are still functioning as rare forest resources at the present.

However, the Olympics also had negative outcomes that resulted from the improvements of the infrastructure that had suddenly become necessary. Especially the construction of the Metropolitan Expressway and the Tokaido Shinkansen were of the highest priority for the country and for Tokyo and improvements in a short time were promoted as procedure. Construction of the former was already demanded before the decision of holding the Olympic Games by the Capital Committee on Construction since 1953 (Table 2), and in the same year the Olympics were decided the Metropolitan Expressway Public Corporation Law was approved (1959 Legislative Act No. 133), making the construction of the expressway an even more urgent matter (Horie, 1996; Furukawa, 2013). Therefore, to avoid time-consuming

Table 2. Rate of properties for Urban Express Highway

Planning agent	Length of Extended Express Highway	River	Public road	Other properties		Total
				Total	Civil properties	
Committee for capital construction	49km	9.10%	7.70%	83.20%	70.20%	100%
Tokyo Metropolitan Government	62.5km	41.80%	34.40%	23.80%	10.90%	100%
Urban planning	71km	35.00%	37.70%	27.30%	14.40%	100%

Source: (Ishihara & Takahashi, 2015)

land purchases, the expressway was built by using the space above state-owned roads, rivers, moats, and waterways (Figure 2). This is also clear from the land use ratio of the expressway, with river land having a ratio of 9.1% (4.9 km) in the planning stage of 1953, increasing to 41.8% (62.5 km) in 1958, and in the process making 35% (71 km) of the used land obtainable without the need to purchase it (Table 2). Among them are planned expressways located between Olympic facilities (National stadiums, Olympic Village) and Haneda Airport such as the Metropolitan Expressway Route 1 Haneda Line (Chuo Ward Honcho 3-Chome – Haneda Airport) and the Route 4 Shinjuku Line (Chuo Ward Nihonbashi Kokucho – Shibuya Ward Yoyogi Hatsudai) with a total of 31.3 kilometers which have been designated as Olympic-related roads and therefore have been partly constructed as bridges above and along canals (Figure 2).

During the era of high economic growth, the relation between human society and the water environment in Tokyo changed remarkably. First, in Japan with its high risk of natural disasters, in the city area mainly flood damage occurred caused by large-scale typhoons, and at the same time, chronic water shortages were caused by rapid urbanization, demanding immediate water management measures. Furthermore, Tokyo has the topographic characteristic that it can be divided into an eastern and a western part. In the lowlands in the eastern part, irrigation canals became unnecessary with the transformation of the agricultural structure. The city's moats as focus of the waterside remnants were filled with earth from construction sites during the reconstruction of war damages and as already stated used for the construction of the Metropolitan Expressway. In the western Yamanote (uptown) area, rivers without headspring were filled up and the construction of sewage and drainage canals was furthered (Tsuboi, 2003). Additionally, especially from the late 1950s until the late

Figure 2. Metropolitan Expressway in Tokyo, Chiyoda Ward
Source: (Photo taken by the author, June 23, 2018)

1960s, the water pollution of small and medium-sized rivers in Tokyo increased due to the inflow of domestic and industrial wastewater, leading to a very low amenity value for the city's inhabitants. Therefore, it was demanded, and requested by the city's residents as well, that small and medium-sized rivers were either to be used as sewage system or covered up (Ishibashi & Takahashi, 2015). For example, the Shibuya River, which is the origin of the place name Shibuya, the world-famous tourist spot, and which is one of the small rivers that flows southwards through the city's center, was covered up in 1961 according to the 36th report of Tokyo and remains in this state until today (Figure 3) (Nakamura & Oki, 2009). For that reason, the number of Tokyo's rivers and waterways has declined after the Meiji era and the water amenity of this period has further decreased with the vanishing of the rivers from the city's surface (Arai 1996).

Tokyo Olympics 2020: Towards Sustainable Development Replacing Developmentalism

Among the activities of the Japanese government to vitalize its Visit Japan Campaign, which was started in 2003, the wish to host the Olympic Games emerged and promotional activities commenced in 2011 and in September 2013, Tokyo was

Figure 3. Riverside landscape in Tokyo, Shibuya Ward
Source: (Photo taken by the author, May 2, 2017)

selected as host city of the 2020 Olympic Games. In the same year, because of the great relaxation of visa requirements for the Southeast Asian countries Thailand and Malaysia (July 2013), a low exchange rate of the Japanese Yen, the expansion of airline routes, and countermeasures to harmful rumors, the number of foreign visitors to Japan reached more than 10 million people, making the importance of the Japanese tourism industry apparent (Hompo & Yagasaki, 2015). With Tokyo's experience as host city of the Olympics, it can make full use of the athletic facilities constructed in 1964 and the infrastructure improvements of that time to keep the environmental impact low and host sustainable Olympics, merits with which the city tried to appeal (Saito, 2014). In fact, in the evaluation reports for each of the candidate cities for the 2020 Olympics—Madrid (Spain), Istanbul (Turkey), and Tokyo—Tokyo received recognition in this respect with a special mention (International Olympic Committee, 2013). The awareness of reducing the environmental impact in hosting the Olympics is set forth in the Olympic Charter of 1994 and contains that it shall not be a big event with environmental sacrifices led by development, since those evaluation standards were specified in the selection points (Ministry of Environment, 2014), protection of the environment and reduction of waste, usage of low-carbon energy, and improvement of sustainability have been consciously followed for every event (Table 3). Especially during the construction of the facilities for the 2012 Summer Olympics in London, evaluation systems for the building environment (e.g. BREEAM) were introduced and ISO standards for event management were created, which were highly appreciated when they became standards for the subsequent Olympics (Ministry of Environment, 2014).

Under such circumstances Tokyo was selected as host for the 2020 Olympics and Paralympics and concerns were directed at environmental maintenance as well as an increased awareness of sustainable development. The concrete environmental policies of the Japanese Ministry of Environment and the Tokyo Metropolitan Government Bureau of Environment at the current stage are summarized in Table 4. With having concluded the Paris Agreement, Tokyo will enact more specific policies with the year 2020 as aim. On the other hand, the Ministry of Environment has indicated that the destruction of the natural environment during the 1964 Olympics, namely the covering of the Shibuya River and the Tachiai River and the construction of the Metropolitan

Table 3. Environmental measurements for Olympics from 1990s

Year	Hosting city	Conservations of natural environment	Prevention of environmental pollution	Waste reduction	Re-use of former built environment	Usage of public transit	Energy saving	Measurement of CO2
1994	Lillehammer	✓		✓				
1996	Atlanta					✓		
1998	Nagano	✓	✓	✓	✓			
2000	Sydney	✓	✓	✓	✓	✓	✓	✓
2002	Salt Lake City	✓	✓	✓				✓
2004	Athens	✓	✓			✓		
2006	Turin	✓	✓	✓			✓	✓
2008	Beijing	✓	✓	✓	✓	✓	✓	✓
2010	Vancouver	✓	✓	✓			✓	✓
2012	London	✓	✓	✓	✓	✓	✓	✓
2014	Sochi	✓	✓	✓	✓		✓	✓

Source: (Ministry of the Environment, 2014)

Expressway above the Nihonbashi River, combined with the increase of high-rise buildings due to the relaxation of the 31m height restriction, are factors that generate Tokyo's heat island phenomenon (Ministry of Environment, 2014). The same report indicates that there is an awareness of the differences to the environmental policies of 1964. Concrete policies are the conservation of existing green spaces, the increase of water permeability of roads and parking spaces and application of water-holding coatings, the laying of electric wiring underground, the maintenance of roadside trees and greening of rooftops and walls of surrounding buildings, the reduction of artificial heat disposal by using high-efficiency equipment, the restoration of water surfaces, the improvement of air circulation of surrounding blocks, with the cool air from rivers and green spaces cited as a prime example, and as measures until the hosting of the Olympic Games, planting of trees around event facilities and along streets show a very locally-oriented approach. Such measures of the governmental environment agencies, it can be said that the developmentalism of the period of rapid economic growth has to no small extent changed in introspective Japan, but on the other hand, for example real estate development by railway companies in Japan, cities were formed after the period of rapid economic growth by leadership of private enterprises (Ministry of Land, Infrastructure, Transport and Tourism, 2011: 82). Accordingly, the next part will look at the projects of private sector entities as projects aimed at the 2020 Tokyo Olympics.

Table 4. Environmental measures towards Tokyo Olympics

Policy subject	Policy objectives	Environmental measures towards Tokyo Olympics	
Ministry of the Environment	2020 Goals/ 2030 Goals/ 2050 Goals	Developing a low-carbon society Mitigation of heat island, heat illness and creation of greenery Realization of clean and comfortable air environment Realization of clean and comfortable aqueous environment Promoting 3R(reduce, reuse, recycle) Developing a sustainable society	
Bureau of Environment	2020 Goals/ 2030 Goals	Climate change and urban energy Sustainable materials and waste management Urban biodiversity and creation of greenery Clean and comfortable air	Sustainable building policy Mitigation of urban heat island Expansion of renewable energy introduction Creation of a hydrogen-based society

Source: (Ministry of the Environment, 2014. & Bureau of Environment, 2017)

PRACTICES TOWARDS SUSTAINABLE CITY IN TOKYO

In Tokyo, private enterprise-led urban redevelopment is gaining momentum at the present due to the estimated growth of the tourism industry in the wake of the hosting of the 2020 Olympics and other events after that, such as the OSAKA Expo, MICE tourism, and integrated resort development (Ikeda, 2017). Among this redevelopment there are many environmentally-friendly projects. This section will provide examples of this eco-friendly redevelopment, such as rooftop and wall greening projects and the renaturation of small and medium-sized rivers by private enterprises, and it will investigate current activities regarding the environment in Tokyo.

Greening Projects

In 2013, the ratio of green space coverage in Tokyo was 19.8% throughout the 23 wards compared to 50.5% in the whole Tokyo Metropolis, showing a decrease by 0.2% since 2008 (Bureau of Environment, 2016). As measure against this decrease of green space coverage 1.01 million roadside trees were planted between 2007 and 2015 (Bureau of Environment, 2016). Furthermore, the "New Development of Green Policies" of 2012 specifies the conservation of existing green spaces, the planting of native species, and the raising of public awareness regarding biodiversity as measures. On basis of the 2014 "Tokyo long-term vision" measures for the conservation of biodiversity and the increase of quality of green spaces, and the 2016 "Tokyo environment basic plan" demands the thorough conservation of the city's environment, showing a formulation of measures regarding Tokyo's green spaces in rapid succession during the 2010s.

The reason that lies behind the development of these policies is the heat island effect caused by the loss of green spaces. Especially Tokyo has a remarkable heat island phenomenon, with a temperature increase of 3.2 degrees centigrade over the past 100 years, which is a rise twice as high as in other 15 urban areas of Japan where the increase of the mean temperature amounts to around 1.5 degrees centigrade (Tsunematsu et al., 2016). In addition, in 2013 the time with temperatures above 30 degrees centigrade, which is related to causing heat strokes, has increased in Tokyo's eastern central wards Chuo, Chiyoda, Taito and Bunkyo by 25% and in other wards by 20%, and the number of days with tropical nights (nights with temperatures not below 25 degrees) now lies between 30 and 40 per year (Tsunematsu, 2014). As a countermeasure, the Tokyo government has revised the "Regulations regarding the conservation and recovery of nature in Tokyo" (from here "Regulations for nature conservation") in April 2001, which require newly constructed or reconstructed buildings with a site area of more than 1,000 square meters (public facilities with more than 250 square meters) to green 20% of the roof's area that can be greened with the corresponding size of the rooftop area (Bureau of Environment, 2016). Furthermore, the "Committee on Countermeasures for the Heat Island Effect in Tokyo" was established in 2002 as well as the "Heat Island Countermeasures Policy Initiative" in 2003. In addition, based on a map of the thermal environment of Tokyo's 23 wards that shows the effects of anthropogenic waste heat and ground surface covering on the atmosphere, which are considered to be causes for the heat island effect, "Areas for the Promotion of Heat Island Countermeasures" were dedicated, comprising four areas (center, Shinjuku, Shinagawa train station area, Osaki-Meguro) (Chiyoda, Shinjuku, Minato, Meguro, and parts of Shinagawa Ward). In these areas, apart from greening projects conducted in cooperation by public (state and ward) and private enterprises, the building environment plan system was adopted which promotes the water retention of roads as well as heat insulation of road surfaces and energy-saving measures for private facilities. In other words, whereas greening projects, including the creation of green tracts of land, were until then under the complete responsibility of the government, private-sector initiatives has started to join measuring heat island effects. However, because of the great differences of environmental circumstances between Tokyo's wards and its outskirts, inner-city and suburban area, it is necessary to consider some local practices related to the Olympics as well. Therefore, in the following the greening projects will be divided into two main categories, the greening of tracts of land and the greening of building rooftops and walls, and examples for projects in the wards that are deemed remarkable by the author will be examined in more detail.[4]

Land Greening Projects and Redevelopment in Chiyoda Ward

Compared to rooftop and wall greening projects land greening requires large plots of land, but large-scale greening projects are difficult to realize because many of Tokyo's redevelopment projects are troubled by complicated ground rights. It is Chiyoda Ward that has shown the most activity in land greening projects. Chiyoda, with Tokyo Station as its hub, is lined with offices and commercial buildings and had the most pronounced difference in daytime and nighttime population during the 1990s. Therefore, compared to the outskirts the temperature drop at night is low and the number of tropical nights is high in comparison to other wards, making it a ward with a considerable heat island effect. For this reason, Chiyoda was chosen as main area of Tokyo's "Heat Island Measures Promotion Area" in 2005. Furthermore, in 2009 it was designated as model block in the pilot project "Cool City Center" which is part of the Ministry of Environment's "Low-carbon Society Model Block Formation Promotion Project".[5] In the case of being acknowledged as "cool city model block", half of the costs of related projects are subsidized. In addition, this project is only intended for private enterprises with corporate status.

To begin with, Chiyoda's total area of green space amounts to 244.91ha and has increased by 8.03ha since 2003, leading to a green space ratio of 21.04% (2010). Due to the Imperial Palace and many parks the area of green space is large, but regarding green space ratio Chiyoda stands at number six among all the wards (Shibuya City, 2016). Chiyoda ward is divided into seven administrative areas of which the Fujimi area with the Imperial Palace at its center has the highest green space coverage (41.45%), followed by the Otemachi area (comprising Otemachi, Marunouchi, Yurakucho, Nagatacho) in the south-east of the Imperial Palace (20.70%). In the Otemachi area, approximately 13 kilometers of streets were greened in a roadside greening project. This corresponds to around 20.5% of all the city block streets in this area and represents the highest ratio among all the administrative areas of Chiyoda ward. The above shows clearly that in Chiyoda Ward the improvement of green spaces is remarkable in the Otemachi area. The change of greening along roads shows an increase of 1.8% in all the wards, but Chiyoda exceeds this clearly with an increase of 4.0% (Chiyoda City, 2010). As reason for this can be seen the establishment of the "Council for the Promotion of a Redevelopment Plan in Otemachi, Marunouchi, and Yurakcho" in 1988, centered around local landowners in the Marunouchi area, which lead to a complete redevelopment on a large scale; and in addition, since this development was lead by private enterprises it was adopted by "the cool city model block".

Rooftop and Wall Greening Projects, Redevelopment in Minato and Shibuya Ward

A major role in green space policies play projects for the greening of rooftops and walls of buildings. The greening of rooftops and walls has been acknowledged as having a relaxation effect on the heat island phenomenon and has been promoted and implemented in laws since the year 2000. The change in construction area for the greening of (building) rooftops and walls in Japan shows for both an increase until 2008, then a temporary decrease in 2009, with the former increasing again slowly after 2013, and the latter increasing slowly as well except for the temporary increase in 2011 (Figure 4). The ratio of construction area for each category according to regions shows that Tokyo Prefecture accounts for 38.8% of all rooftop greenings in 2016, followed by the neighboring prefectures Kanagawa with 9.4% and Saitama with 5.2%, making the Tokyo Metropolitan Area account for a total of 53.4% throughout Japan. Apart from that, the prefectures Aichi (8.7%), Osaka (7.9%) and Hyogo (6.7%) show nationwide high numbers as well. Regarding wall greening, the same prefectures as before are occupying the first six ranks, with the Tokyo Metropolitan Area accounting for 53.3%. However, Aichi prefecture (12.3%) and the Kinki Metropolitan Area's prefectures Osaka (9.6%) and Hyogo (7.4%) show slightly higher numbers in this category. In other words, most of the rooftop and wall greening projects can be found in and around Japan's three largest cities where the heat island phenomenon is most striking where the city size is the biggest and the number if construction starts is high. In a similar way, the change in greened area of rooftops in the Tokyo Metropolis shows a steady growth in the first half of the 2000s, and a decline in the latter half; from 2010 until 2012 as well as from 2015 on, an increase can be observed (Figure 4). The decline in the latter half of the 2000s can be explained with the economic stagnation caused by the global financial crisis in 2008 which led to a reduction in the number of new building constructions.

In their master plans for parks and open spaces in which each local government establishes its rooftop greening projects, the only wards that have formulated specific goals are Taito (in 2002), Shibuya (2004), Suginami and Kita (both 2010) (Takeuchi, 2012). From this can be inferred that rooftop greening plays an important role in the greening projects of these wards. However, the wards with the largest green areas on rooftops are Minato (12.7ha), Shibuya (11.4ha) and Setagaya (10.2ha) (Nakabayashi, Kishii, Oosawa, 2012). In the following part, examples from the two wards occupying the first two places, Minato and Shibuya, will be examined.

The total area of green space in Minato Ward amounts to 452.26ha, corresponding to a relatively high green coverage ratio of about 21.8%. Among this total area, the ratio of public green spaces is 58.9%, whereas the ratio of public spaces among all greened rooftops is low with 18.7%, and 85.6% of all greened rooftops are privately owned (Minato City, 2017). This data supports that rooftop greening is mainly utilized by private enterprises. Among the five areas in this ward—Shiba, Azabu, Akasaka, Takanawa, Shibaura Konan—green coverage is the highest in Akasaka with 31.7% due to several shrines and the Aoyama Cemetery. In contrast, the highest number of greened rooftops has Azabu with 168.5 places per square kilometer, and

Figure 4. Transition in the area of rooftop and wall greening from 2000 until 2016 in Metropolitan areas
Source: (Ministry of Land, Infrastructure, Transport and Tourism, 2017)

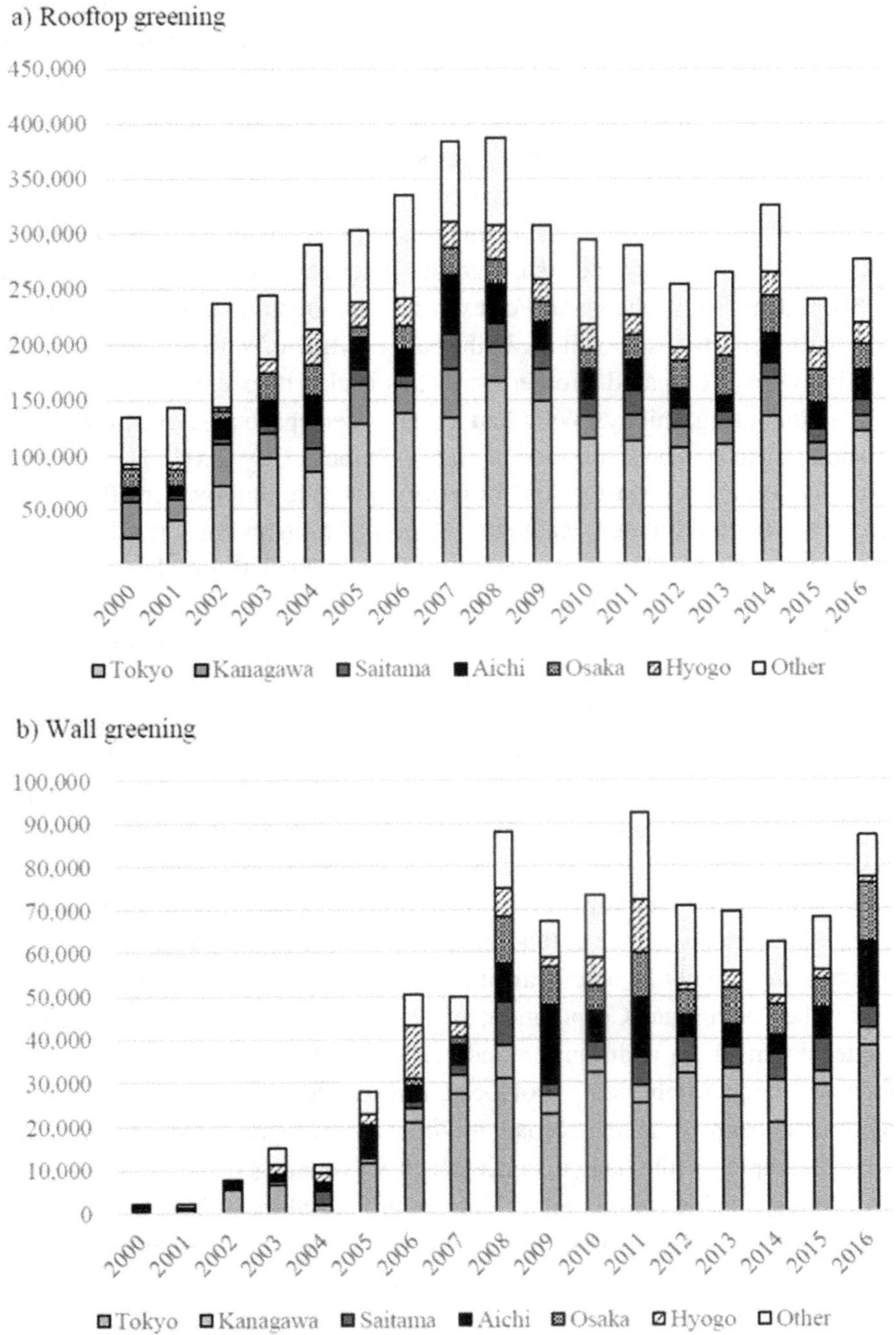

wall greening is highest in Shiba with 95.7 places per square kilometer. The Azabu area is home to the world-famous Roppongi Station and in the 1980s land owners and the developer Mori Building Company carried out a large-scale redevelopment of the whole area. Ark Hills, which is located between Minato's Akasaka 1-chome and Roppongi 1-chome, is a high-rise building of 153 meters height completed in 1986 that comprises offices, accommodations, high-class apartments, cultural and media facilities. At this corner, the Ark Garden was created which was a pioneer of rooftop gardens (specifically, hanging gardens) by a private enterprise before the promotion of rooftop greening was started.

On the other hand, with plans for the main sports stadium of the Olympics to be located in Minato's bayside area, vigorous practices related to greening projects have been identified in the ward. For example, the greening plan devised by each ward and the reports based on it are in the case of Minato Ward provided in general every five years. Compared to other wards this is clearly extensive.

The neighboring Shibuya Ward had a total green space area of 311ha in 2003, corresponding to a green space ratio of 20.6% (Shibuya City, 2004). However, since the formulation of the "Master Plan for Green Spaces in Shibuya" in 2004, rooftop greening has been promoted, leading to its propagation and as a result increasing the green space ratio to 21% (Shibuya City, 2016). In 2003, the total area of greened rooftops amounted to 1.35ha, representing less than 1% of the ward's green space coverage (Shibuya City, 2004). The ward is divided into six administrative areas (Honmachi and Sasazuka; Hatsudai, Nishihara and Uehara; Yoyogi; Sendagaya and Jingumae; Omukai and Ebisu; Hikawa and Shinbashi), and with one part of the Shinjuku Imperial Garden, Yoyogi Park and the Meiji Shrine the areas Yoyogi and Sendagaya-Jingumae are the areas with the highest green space ratio among them. Regarding green spaces on rooftops, the largest coverage with 0.88ha has the Omukai-Ebisu area, followed by the Hikawa-Shinbashi area with 0.55ha and the Sendagaya-Jingumae area with 0.52ha. The Omukai-Ebisu area and the Sendagaya-Jingumae area are the center of redevelopment in Shibuya aimed at the Olympics. Both areas are directly located near Shibuya Station, and with the train station at its center the Tokyu Land Corporation, owner of large plots of land, is carrying out the redevelopment. In addition, around Harajuku Station (Sendagaya- Jingumae area), a well-known sightseeing spot located close to Shibuya Station and visited by many tourists from Japan and abroad, multiple redevelopment projects are carried out. For example, at the Tokyu Plaza which was managed by the same private enterprise responsible for the redevelopment around Shibuya Station—the Tokyu Land Corporation—a green space for recreation named "Omohara Forest" was established for its renewal opening in 2016 (Figure 5).

Figure 5. Rooftop garden Omohara-no-mori (Omohara Forest) in Harajuku, Shibuya Ward.
Source: (Photo taken by the author, May 2, 2017)

River Projects

As representative river project for the Olympics the improvement of the hydrophilicity of the Shibuya River—which has shaped the topography of Shibuya (ya of Shibuya 渋谷 means a valley) Station—is pointed out. The Shibuya River is with its total length of 6.6 kilometers one of the small and medium-sized rivers that was diverted into a culvert for use as sewage canal after the "36 Reports" of 1961. The culvert construction was carried out at different times for different parts of the river. The stretch between Yotsuya and Miyashita Bridge (below Shibuya Station) was constructed after 1965 and the part from Miyashita Bridge until Inari Bridge after 1971. The decision to improve the environment of the culverted Shibuya River was made in June 2013, when the hosting of the Olympics was still undetermined (Miki, 2017).[6]

The river projects are not independent projects but are part of the large-scale redevelopment plans for the station area. The planning for this large-scale redevelopment around Shibuya Station began in 2007 and since 2008 urban regeneration special areas (from here on called special areas) were established/specified; 21 areas (districts/zones) in Shibuya (Station) 2-chome in 2008; 21 areas around Shibuya Station, Dogenzaka and 3-chome in 2013; 1 area in Sakuragaokacho in 2014; 15 areas in Udagawacho. [7]

The most important characteristic of the special areas is that for the relaxation of regulations regarding the floor area ratio there is no set standard for the required public contribution, but the urban planning decision holders evaluate the public contribution individually and can make suitable relaxations of regulations. According to Yamauchi, Okata, Koizumi, Murayama, Manabe (2015), there are nine kinds of public contributions in the Shibuya Station area, including the offering of spaces for people who are unable to return home directly because of a natural disaster, the city functions that increase international competitiveness and the environmental improvement of the riverfront by renaturing the Shibuya River and promoting three-dimensional greening. For these reasons, Shibuya Ward and Tokyu Land Corporation work in collaboration on these projects, while conserving and restoring the green of the Shibuya River, utilizing a landscape shaped by a waterfront and green spaces, be the goals of these projects. These projects are ongoing at the present and after completion they are planned to provide a green promenade of approximately 600 meters and two public squares (Tokyu Corporation, 2013). Furthermore, not only parts of the Shibuya River's environment but also its upper reach and the green way along the former waterway of the Tama River waterworks are considered for improvements in the future (Shibuya Ward, 2016).

DEVELOPING A SUSTAINABLE ECO-CITY IN PRE-OLYMPIC TOKYO: POTENTIAL OF NEW METHODS AND THEIR LIMITS IN AN URBAN ERA

This chapter reflected about the environmental sustainable city in the urban era by examining the example of the world's most populous city, Tokyo. The city, which will be hosting the Olympic Games for a second time in 2020, has taken environmental problems consciously into consideration since the bidding process for this mega event. Particularly during the 1964 Olympics, when Japan was in a reconstruction period after the war, the city took a developmentalist approach regarding the improvement of its infrastructure, and especially the rivers and waterways disappeared in this process.

Through introspection about this developmentalism, the administrations of Tokyo and Japan came up with an environmental policy that emphasized a more sustainable urban development. As examples for this, greening projects and the improvement of the waterside environment were given in this chapter. As measures against the heat island effect in Tokyo's inner city, the creation of green spaces was advocated by the metropolitan and the national administration through the enactment of several laws since the 2000s. Propelled by this improved legal situation, land greening projects (Chiyoda Ward) as well as rooftop and wall greening projects (Minato Ward, Shibuya Ward) by private enterprises started to gain momentum throughout Tokyo's 23 wards. In addition, the projects for improving the waterside environment (Shibuya Ward) aim at increasing the amenity value of parts of the city's small and medium-sized rivers that have vanished from the cityscape during Japan's period of rapid economic growth. Private companies are the main actors practicing greening projects, utilizing the administrative legal framework in the pre-Olympic resume, which promotes fundamental urban restructuring as an opportunity.

However, it is uncertain whether these efforts lead to a fundamental solution of the urban heat island issue. Therefore, further research and investigation are eagerly awaited. In addition, as part of the promotion of post-Olympic urban tourism, the promotion of the nighttime economy and night-time activities is currently getting attention (Ikeda, 2017). This could result in the new environmental burden for the city's environment. In this regard, future research is required.

ACKNOWLEDGMENT

This research was supported by the JSPS [grant numbers 17J02079]; the Sasakawa Sports Foundation [grant number 180A2-032].

REFERENCES

Aikawa, M., Hiraki, T., & Eiho, J. (2006). Vertical atmospheric structure estimated by heat island intensity and temporal variations of methane concentrations in ambient air in an urban area in Japan. *Atmospheric Environment*, *40*(23), 4308–4315. doi:10.1016/j.atmosenv.2006.03.044

Arai, T. (1996). Changes in the hydrological environment in Tokyo. *The Journal of Geography*, *105*(4), 459–474. doi:10.5026/jgeography.105.4_459

Bornstein, R. D. (1968). Observations of the urban heat island effect in New York City. *Journal of Applied Meteorology and Climatology*, *7*(4), 575–582. doi:10.1175/1520-0450(1968)007<0575:OOTUHI>2.0.CO;2

Bureau of Environment. (2016). *Tōkyō-to kankyō hakusho 2016* [White paper on the environment of Tokyo 2016]. Retrieved from http://www.kankyo.metro.tokyo.jp/basic/plan/white_paper/2016.html

Butler, T. (2015). *Overdevelopment, overpopulation, overshoot*. New York, NY: ORO Editions/Goff-Bools.

Chiyoda City. (2010). *Chiyoda-ku midori no jittai chōsa oyobi netsu bunpu chōsa gyōmu hōkokusho* [Report on green space and distribution of heat issues in Chiyoda Ward]. Retrieved from https://www.city.chiyoda.lg.jp/koho/machizukuri/kankyo/ryokuka/heat-island.html

Echizawa, A. (2014). 1964-nen Tōkyō orinpikku to toshi keikaku [1964 Tokyo Olympics and city planning]. *Urban Housing Sciences*, *87*, 24–28.

Furukawa, K. (2013). *Shuto kōsoku dōro no saisei* [Renewal of Metropolitan Expressway]. Reference.

Hompo, Y., & Yagasaki, N. (2015). Japan's inbound strategies using 2020 Tokyo Olympic and Paralympic Games based on the experiences of the past games [in Japanese]. *The International Journal of Tourism Science, 8*, 3-11.

Horie, K. (1996). Planning process of expressway in Tokyo [in Japanese]. *Doboku Keikakugaku Kenkyu, Ronbunshu, 13*, 1–22. doi:10.2208/journalip.13.1

Ikeda, M. (2017). A review of nightlife literature and suggestions for future research in Japan. *Geographical Space*, *10*(2), 67–84.

International Olympic Committee. (2013). *Report of the 2020 evaluation commission*. Retrieved from https://stillmed.olympic.org/Documents/Host_city_elections/2020_Evaluation_Commission_report.pdf

Ishibashi, N., & Takahashi, T. (2015). *Background of the exclusive use of the elevated Metropolitan Expressway over Nihonbashi River*. The Institute of Civil Engineering of the Tokyo Metropolitan Government Annual Report.

Kondo, H. (2009). City canopy [in Japanese]. *Tenki*, *56*(8), 677–678.

Lee, J., Harashima, Y., Lee, D. K., & Morita, T. (1995). A comparative analysis of processes used in developing environmental policies in Japan and Korea. *Environmental Sciences*, *8*(2), 181–192.

Maeda, C., Kondo, A., Akikawa, M., Kaga, A., & Inoue, Y. (2008, March): *Simulation of urban heat island in Hyogo*. Paper presented at the meeting of the Society of Heating, Air-conditioning Sanitary Engineers of Japan, Osaka, Japan.

Maruyama, M., & Nakagawa, Y. (2004). Shutoken seibi keikaku ni okeru ryokuchi seisaku no hensen ni kansuru kisoteki kousatsu [Basic consideration about the transition of green policies in Metropolitan Region development]. In *Proceedings of the Meeting of Japan Society of Civil Engineers* (Vol. 59, pp. 4-334). Tokyo, Japan: Japan Society of Civil Engineers.

Masumoto, K. (2007). Space-time characteristics of air temperature distribution and heat island situation in Osaka City. *Journal of Urban Living and Health Association*, *51*(4), 215–224.

Matsuoka, R. (2017). *Toshi ni okeru ryokuchi ni kansuru seido keikaku no hensen to Tōkyō 23-ku no ryokuchi seisaku no tokuchō* [Transition of institution and planning related to urban green spaces and the features of green policy in Tokyo's 23 wards] (Unpublished graduation thesis). Tokyo, Japan: Waseda University.

Miki, T. (2017). Redevelopment projects in Shibuya Tokyu Corporation [in Japanese]. *Journal of JSEE*, *65*(6), 41–43. doi:10.4307/jsee.65.4_1

Minato City. (2017). *Minato-ku midori no jittai chōsa* [Report of Green Investigation in Minato]. Retrieved from https://www.city.minato.tokyo.jp/ryokukasuishin/kankyo-machi/kankyo/chosa/9midorinojittai.html

Ministry of Land, Infrastructure, Transport and Tourism. (2011). *Shuto-ken seibi hou ni motozuku daitoshi-ken no minaoshi* [Re-examining the Tokyo metropolitan regional policy based on Metropolitan Region Development Law]. Retrieved from http://www.mlit.go.jp/common/000139692.pdf

Ministry of Land, Infrastructure, Transport and Tourism. (2017). Zenkoku okujō hekimen ryokuka sekō jisseki chōsa no kekka hōkoku [Report of investigation of rooftop and wall greening in Japan] [Data file]. Retrieved from http://www.mlit.go.jp/report/press/toshi10_hh_000257.html

Ministry of the Environment. (2014). *2020-nen orinpikku pararinpikku Tōkyō taikai wo keiki toshita kankyō hairyo no suishin ni tsuite* [Report on promoting the environment policy for the 2020 Olympics and Paralympics]. Retrieved from https://www.env.go.jp/press/18532.html

Mizuno, M. (1993). Subjects on environment in 24-hour city: Discussion on urban heat island [in Japanese]. *The Journal of the Acoustical Society of Japan*, *49*(11), 832–838.

Nakabayashi, S., Kishī, T., & Ōsawa, M. (2012). On the rooftop gardening policy in the 23 wards of Tokyo: Focused on the actual condition of the public buildings [in Japanese]. *Journal of the City Planning Institute of Japan*, *47*(3), 475–480.

Nakamura, S., & Oki, T. (2009). The philosophy and chronicle of abolished urban river in 36'report [in Japanese]. *Proceedings of Hydraulic Engineering, JSCE*, *53*, 565–570.

Oke, T. R. (1979). Review of urban climatology 1973-1976. *WMO-Technical Note*, *169*, 1–100.

Ozaki, M. (2002). A Study of the Process of Forming Sport Policy in Japan [in Japanese]. *Hitotsubashi University Research Series Humanities*, *39*, 159–252.

Saito, T. (2014). Orinpikku pararinpikku to kankyō risku no kanri [Olympics, Paralympics and the management of environmental risks]. *Sompo Japan Nipponkoa Risk Management Report*, *126*, 1–11.

Sakakibara, Y. (2001). Comparison between the effect of heating from urban surface and that of mechanical mixing of urban atmosphere to heat island [in Japanese]. *Tenki*, *48*(5), 305–311.

Shibuya City. (2004). *Shizen kankyō chōsa hōkokusho* [Report on the investigation of the environment]. Retrieved from https://www.lib.city.shibuya.tokyo.jp/?action=common_download_main&upload_id=514

Shibuya City. (2016). *Shibuya-ku midori no seibi hōshin* [Green policy in Shibuya Ward]. Retrieved from https://www.city.shibuya.tokyo.jp/kusei/shisaku/ku_keikaku/green_plan.html

Takeuchi, K., & Lee, D. K. (1988). A framework for environmental management planning [in Japanese]. *Journal of the Japanese Institute of Landscape Architects*, *52*(2), 95–104. doi:10.5632/jila1934.52.95

Takeuchi, T. (2012). A study on the potential of Green Structure Plan as comprehensive spatial control plan [in Japanese]. *Journal of the Japanese Institute of Landscape Architecture*, *75*(5), 601–604. doi:10.5632/jila.75.601

Takeuchi, T., & Ishikawa, M. (2008). A study on the green space policies on the fringe of Tokyo wards area in the 1950's and 60's [in Japanese]. *Journal of the City Planning Institute of Japan*, *43*(3), 199–204.

Tokyo Metropolitan Government. (2017). *New Tokyo. New tomorrow. The action plan for 2020.* Retrieved from http://www.metro.tokyo.jp/english/about/plan/index.html

Tokyu Corporation. (2013). *Shibuya-eki minami gaiku purojekkuto ni kansuru toshi keikaku no kettei ni tsuite* [Notification of the decision about the south district project at Shibuya Station]. Retrieved from https://www.tokyu.co.jp/file/130123-3.pdf

Tsuboi, S. (2003). A study on change of water area and water activity projects in Tokyo ward area [in Japanese]. *Comprehensive Urban Studies*, *82*, 19–34.

Tsunematsu, N. (2014). *Tōkyō ni okeru hiito airando no jittai to shonetsu taisaku ni tsuite* [Heat island in Tokyo and countermeasures against heat]. Retrieved from https://www.tokyokankyo.jp/kankyoken/wp-content/uploads/sites/3/2014/12/4cc271adc7839d797e861f8940f5e06c.pdf

Tsunematsu, N., Yokoyama, H., Honjo, T., Ichihashi, A., Ando, H., Yamagata, Y., Murakami, D., & Shigyo, N. (2016). Relationship between land use variations and spatiotemporal changes in amounts of upward infrared radiative energy in downtown Tokyo at midday on hot summer days. *Annual report of the Tokyo Metropolitan Research Institute for Environmental Protection*, 76-82.

UN Habitat. (2016). *World cities report 2016: Urbanization and development: emerging futures*. Retrieved from http://wcr.unhabitat.org/main-report

United Nations. (2016). *The World's Cities in 2016 Data Booklet*. Retrieved from http://www.un.org/en/development/desa/population/publications/pdf/urbanization/the_worlds_cities_in_2016_data_booklet.pdf

Uno, I., Wakamatsu, S., & Ueda, H. (1988). Behaviour of nocturnal urban boundary layer and air pollutants [in Japanese]. *Journal of Japan Society of Air Pollution*, *23*(2), 103–114.

Yamauchi, T., Okata, J., Koizumi, H., Murayama, A., & Manabe, R. (2015). A study of the process to determine public contribution elements in spatial districts for urban generation – Case studies in Ochanomizu, Shibuya, Ginza districts [in Japanese]. *Journal of the City Planning Institute of Japan*, *50*(3), 904–911.

ENDNOTES

1 The general explanation for this effect is that bulks of hot air rise from the surface and expand, and as a result the difference to the ambient air temperature disappears leading to a stagnation at a constant altitude, forming the dome of urban air (Mizuno, 1993).

2 The Basic Law for Environmental Pollution Control was changed into the Basic Environment Law in 1993 and has specified the principal policy of Japan's environmental administration since then.

3 These projects were carried out by private railroad companies during the development of the railway lines around Tokyo and consisted mainly of housing land development for company managers and senior bureaucrats.

4 In addition, actions regarding greening projects in the wards are not easy to compare merely by their numbers. Matsuoka (2016) compares the public and private practices for each greening category and shows that projects for the greening of rooftops are also easy to handle for private enterprises and that in the wards Chiyoda, Koto and Edogawa public greening projects and in the wards Minato and Itabashi private projects are most active.

5 The criteria for the selection as model block are a significant heat island phenomenon, the easy attraction of general attention with a high effectiveness to promote further spreading, and the high probability to enforce concentrated measures.

6 However, according to developer Tokyu Land Corporation there was the formation of a consensus saying that "The city functions around Shibuya that were shaped during the last Olympics in 1964 have not much changed since then and the people concerned had a common understanding that a reconstruction was necessary", which suggests an awareness that the city's structure and functions were determined by the Olympics (Miki, 2017).

7 These areas are, based on the Urban Renaissance Special Measure Law enacted in 2002, in the areas for city regeneration urgent improvement based on the existing zoning, excluded from regulations regarding usage and floor area ratio, making a high flexibility in planning possible.

Chapter 10
Forest-River-Ocean Nexus-Based Education for Community Development:
Aiming at Resilient Sustainable Society

Shimon Mizutani
Tokyo University of Marine Science and Technology, Japan

Kai Liao
Tokyo University of Marine Science and Technology, Japan

Tsuyoshi Goto Sasaki
Tokyo University of Marine Science and Technology, Japan

ABSTRACT

Bioeconomic research aims at developing a more resource-efficient and sustainable society that uses renewable biological resources to produce food, materials, and energy. Economic supremacy causes many problems, such as global warming, depletion of fossil fuels and natural resources, and loss of biodiversity. In order to build a more sustainable society with resource efficiency, it is necessary to discuss the institutional framework, which includes environmental assessment, environmental monitoring, biological resource management, human resources management, and education. This chapter examined the effectiveness of forest-river-ocean nexus-based education for community development (FRONE) in encouraging the sustainable use of biological resources. Combined with the adaptive cycle, FRONE is considered to have the potential to promote the sustainable use of biological resources. In the future, further bioeconomic research from the point of view of the education system will be needed.

DOI: 10.4018/978-1-5225-7958-8.ch010

INTRODUCTION

Bioeconomy (BE) is characterized by the creation and efficient use of natural and biological resources, raw materials, and capabilities in sustainable infrastructures, which are aimed at, for example, the bio production of goods, bio services, bio energy, and bio health, to achieve sustainable lifestyles, wealth, and economic growth (Hernández, Pallagst & Hammer, 2017). As one of strategies of BE, rural development is a key issue. In order to accomplish this purpose, sustainability of production should occur in the country of origin (FAO, 2016).

The authors focused on local development using local biomass connected to local people's daily life and memories depending on local environment. They examined the effectiveness of forest-river-ocean nexus-based education for community development (FRONE) in enhancing the sustainability of the uses of biological resources. FRONE is an educational program where cherry salmon was used as the teaching tool, because it symbolizes the forest-river-ocean nexus. Based on the learning of scientific and local knowledge about this species, participants will be able to understand the fact that the connection between human and forest-river-ocean nexus exists. The development of FRONE can be divided into five phases, as a result of the Great East Japan Earthquake. Evidence shows that, when FRONE is combined with adaptive cycle theory, FRONE can contribute to the development of a resilient and sustainable community. In other words, FRONE is considered to have the potential to promote the sustainable use of biological resources. In this article, the authors proposed the direction of environmental education to empathize local environment and bio resources in the tsunami devastated area in northern Japan, in order to create a resilient and sustainable society.

BACKGROUND

Japan is an island country that includes approximately 6800 islands, and has a privileged natural environment with bountiful forest, rivers, and ocean. Traditionally, people living in the islands have taken care of these natural environments over generations. Therefore, biodiversity has been maintained by traditional farming societies, forest societies, and fishing communities through a sustainable way of maintenance and the utilization of biological resources. This tradition has also benefited the Japanese society since ancient times. People could live a stable life without worrying about resource depletion, which laid the foundation for Japan's subsequent cultural development.

However, after the high-growth period from the mid-1950s to the early 1970s, its inhabitants' awareness of the coexistence of nature and human beings has been seriously diluted. Dilution of the consciousness caused various environmental problems concerning the connection between nature and human beings (Ministry of the Environment, 2016). Some examples are widespread abandonment of farmland, decrease of animals and plants species in Satoyama, water pollution, and severe climate change.

In order to solve these problems, some fishermen and local residents in northern Japan worked together to make people re-recognize the forest-river-ocean nexus (Wakana, 2001). In 2016, the Japanese Ministry of the Environment announced proposals on the nexus between forest-river-ocean and human beings. The proposal stated that the connection is essential to the establishment of a sustainable society. Such society is featured with environmental, social, and economic sustainability, low-carbon, resource-circulating, and natural symbiosis.

It is well known that the Great East Japan Earthquake caused enormous damage to coastal areas. Since then, eight years have passed, and construction of coastal levees and housing area are still underway. Besides, other problems should not be neglected, such as depopulation and aging (Yamaguchi, 2009), shortage of heirs (Ogaki & Saio, 2014), population decrease (Ministry of Internal Affairs and Communications, 2015), and discrepancy between government and citizen's opinion (Hirata, 2013). Ooe (2014) advocated the necessity of "endogenous reconstruction," in which local people utilize their capacities and contribute to the reconstruction voluntarily. How to realize endogenous reconstruction? Ooe pointed out that to accomplish the restoration and reconstruction of a community-based society, the combination of intrinsic and external power, and the inheritance of traditional and local knowledge are essential. This means to cherish the relationship between people and people, and people and nature.

It is necessary to construct a resilient and sustainable society through an endogenous reconstruction that values the connection between people and nature. The area the tsunami affected is blessed with the natural environment of the forest-river-ocean nexus, where the water of many clean rivers originates in mountains such as Kitakami highland and flows into the sea. Educational practice research in the Hei River watershed in Iwate prefecture was not only focusing on the endogenous development of watershed (including the head stream area, the estuary area, and the coastal area), but also the endogenous reconstruction of the coastal disaster area by implementing FRONE. This program has been taken up in Japan and other countries as a proposal rebuilding a resilient and sustainable society and enhancing the participants' awareness towards the forest-river-ocean nexus in the disaster area (Association for Resilience Japan, 2016; UNESCO, 2017).

MAIN FOCUS OF THE CHAPTER

In this research, the authors analyzed the implementation of FRONE in Miyako city, Iwate prefecture, from educational practical activities in 1994 to today's initiatives. Until now, at the meeting for the development of the environmental education program aiming at endogenous recovery, with the application of Learning Cycle Theory, the dialogue process has established a harmonic relationship, and intrinsic motivation and creativity (Sasaki, 2016).

FRONE was divided into five phases, based on the theory of focusing on the exploration target (FET). The contents of the engagements and outcomes in the five phases were discussed. The five phases are:

Phase 1: Engagements before the Great Earthquake.
Phase 2: Interruption of aquatic marine environmental education (AMEE) by the Great Earthquake.
Phase 3: Restart of AMEE six months after the Earthquake.
Phase 4: Development of AMEE aiming at endogenous reconstruction.
Phase 5: Research results and development of FRONE.

This study also evaluated the resilience of the society by applying the adaptive cycle to confirm whether endogenous reconstruction can help in the construction of resilient and sustainable society in coastal affected areas.

The application of the adaptive cycle to each phase reveals that "reorganization" and "re-recognition" of the nexuses including people and people, and people and forests, rivers, and the ocean were achieved. These results developed in an interrelated way. In each phase, participants acted under universal values, obtained a sense of self-efficacy, and enhanced skills, networks of human relations, and interactive trust. This corresponds to autonomy, competence, and relatedness of the self-determination theory. It is likely that the education program can further stimulate the participants' creativity and intrinsic motivation.

In the fifth phase, the food intrinsic value (FIV) was defined. As the basis of human life, food is derived from the environment of the spatiotemporal forest-river-ocean nexus. A healthy food chain can strengthen people's connection to forests, rivers, and the ocean. The FIV is the recognition that people live in the spatiotemporal connection of forests, rivers, and the ocean through food.

Deepening the recognition and understanding the FIV through the practice of FRONE creates skills, networks of human relations, and interactive trust throughout the watershed region, and leads to the creation of a resilient and sustainable society.

Resilience and Adaptive Cycle

The Concept of Resilience and Adaptive Cycle in Ecological Research

The concept of resilience was first introduced by the Canadian ecologist Holling (1973). He defined resilience as a measure of the ability of an ecosystem to return to its previous natural state after a disturbance. Nowadays, resilience is usually defined as the ability of the ecosystem to absorb the change without shifting to an alternative state, or the ability to maintain the original function and structure during a disturbance (Holling, 1973; Gunderson and Holling, 2002; Walker, Holling, Carpenter, Kinzig, 2004). Despite its roots in ecology, resilience has also been applied to social-ecological systems (Resilience Alliance, 2009).

In order to assess resilience, the adaptive cycle was widely adopted (Resilience Alliance, 2009). An adaptive cycle is proposed as a fundamental unit for understanding complex systems, from cells to ecosystems and societies. It consists of four phases:

1. Growth or exploitation (r).
2. Conservation (K) of established patterns and resource distribution.
3. Collapse or release (Ω).
4. Reorganization (α).

The shift from growth (r) to conservation (K) is referred to as the "fore-loop". It describes the slow and often longer phase of growth and accumulation of resources in the system. The "back-loop" is a rapid transition from release (Ω) to reorganization (α).

Taking the ecosystem as an example to illustrate the adaptive cycle, during the slow sequence from growth or exploitation (r) to conservation (K), interconnectedness and stability increase, and a capital of nutrients and biomass is slowly accumulated. Such interconnectedness can diverse the utilization of space, and has effects on other ecosystems and the next generation. Moreover, the interconnectedness of the ecosystem can facilitate the transition from release (Ω) to reorganization (α). The case of forest fire can be used as an ecological example of the adaptive cycle. Forest ecosystems can take several decades or centuries to move from growth or exploitation (r) to conservation (K) until the system is mature. Following the fire, the system is colonized by different species, such as grass or shrubs. The survival of plant seeds in the soil is a form of natural capital that allows the regeneration of certain original species. This phase is called reorganization (α).

The Application for Adaptive Cycle in Social-Ecological Systems

The development of adaptive co-management of a wetland landscape in southern Sweden provides a good example to understand the dynamics of social-economic system. In the 1980s, due to neglect, inadequate management practices, and decrease in the residents' perception, the wetland ecosystem collapsed. At that time, the steward played a particularly significant role in creating and shaping the organization change, he is also the director of the Ecomuseum Kristianstads Vattenrike (EKV) (Olsson, Folke & Hahn. 2004). In response to ecological changes, four essential processes occur:

1. Integrating knowledge.
2. Developing a vision and goals within a common framework.
3. Developing a robust social network.
4. Recognizing and acting when a window of opportunity opens.

After applying the adaptive cycle model to this case, Olsson, et al. (2004) concluded that during the release (Ω), declining bird population, eutrophication, and overgrowth of the lake occurred. Subsequently, the EKV director started up the reorganization (α). He compiled existing ecological knowledge and experience he found within the network in a project proposal, promoted a management within EKV, which treats humans as part of ecosystems and includes social, economic, and ecological dimensions. This is the first of the abovementioned processes (i.e., integrating knowledge). Then, the regional activist provided overall goals and vision to wetland management. This is the second process (i.e., developing a vision and goals within a common framework). The first two processes belong to the growth or exploitation (r). In order to achieve the new vision and goals, he established a social network based on trust and dialogue. This is the third process (i.e., developing a robust social network). Also, he made an overall plan that considered social, economic, and environmental aspects, and he developed and realized the idea of the EKV. This stage is true for the forth process (i.e., recognizing and acting when a window of opportunity opens). The last two processes could be classified into conservation (K). As the authors described above, it is possible to identify the four phases of the adaptive cycle in the social-ecological system. In particular, for evaluating the resilience of the system, it is important to build skills, networks of human relations, interactive trust on the premise that release (Ω) and reorganization (α) are followed by conservation (K).

Development of Forest-River-Ocean Nexus Based Education for Community Development

The origin of FRONE is a fish ecological investigation conducted in 1994 and continued until 2006. In 2007, the investigation became a part of AMEE, the purpose of this education program is to promote the understanding of the connection between aquatic marine environment and people. AMEE was selected under the Support Program for Contemporary Educational Needs, which was funded by the Ministry of Education, Culture, Sports, Science, and Technology, for promoting aquatic marine environmental literacy education. In 2008, the SANRIKU Education for Sustainable Development (ESD) Hei River Academy (HRA) was established to carry out the selected education program. After the 2011 Great East Japan Earthquake, the education program was developed into FRONE, which still exists today.

This research applied the FET. This theory assumes that an interactive interchange of information, ideas, and knowledge related to a specified exploration target among participants on an equal footing, brings new ideas and outcomes are (Waki, Sasaki & Oshima, 2015). Based on FET, this research divided the development of FRONE into five phases:

Phase 1: Engagements before the Great Earthquake.
Phase 2: Interruption of AMEE by the Great Earthquake.
Phase 3: Restart of AMEE six months after the earthquake.
Phase 4: Development of AMEE aiming for endogenous reconstruction.
Phase 5: Research results and development of FRONE.

Next, the authors clarified the contents and results of each phase and analyzed their significance. Furthermore, they considered these five phases, based on the adaptive cycle model that evaluates resilience, and explored the capability of the creation of a resilient and sustainable society in coastal affected areas by the FRONE.

Phase 1: Engagements Before the Great Earthquake

1. **The Fish Ecological Investigation Conducted by High School Students:** The origin of FRONE is the fish ecological investigation activity. Which was actually the "integrated practice" course for students in prefectural fisheries high school. The study area was the mouth of the Hei River, Miyako city, Iwate prefecture, in 1994. In this activity, they discovered wakasagi (*hypomesus*

nipponensis), which is an important edible fish native to the lakes and estuaries of northern Japan. It triggered their further investigation in the distribution of other fish species, and the life history of wakasagi (Sasaki, 2006a).

2. **Aquatic Marine Environmental Education:** On the basis of the above field survey, the research on AMEE began in 2006 (Sasaki, 2006b). AMEE is based on the theory of the learning cycle. As defined in AMEE, an aquatic environmentally literate person can observe his/her surrounding aquatic marine environment in a scientific way, inquire the related environmental problems and the well-being of human communities, grasp the required comprehensive knowledge (e.g., aquatic marine environmental literacy), make responsible decisions and take responsible actions based on his/her outdoor learning experiences and comprehensive knowledge, and have the capacity to convey them to as many people as possible, effectively. The research outcomes confirmed that the implementation of the AMEE program not only prompts learners' active learning, but also inspires participants' ownership and creativity (Waki et al., 2015).

The basic principle of aquatic marine environmental literacy is grasping the holistic understanding of the connection between aquatic marine environment and people. Such knowledge includes both scientific knowledge (science technology and school education) and local knowledge (traditional ecological knowledge) (Sasaki, 2011a). Local knowledge developed over time by people living in a given community, and is continuously developing. It includes practical experience and skills related to the living animals and the surrounding environment (Research Institute for Humanity and Nature, 2015). Sustainable development requires taking long-term perspectives, integrating local and regional effects of global change into the development process, and using the best scientific and traditional knowledge available. It also requires eliminating the communication gap among the public and achieving better communication (United Nations, 1992). Therefore, it can be expected that AMEE can help to realize a sustainable society.

3. **SANRIKU ESD Hei River Academy (HRA):** In order to protect the rich natural environment of the local community for the next generation, in 2008, in the Hei River watershed of the city of Miyako, Iwate prefecture, watershed residents (e.g., stakeholders of fisheries cooperatives, processors of marine product, retailers, housewives, school teachers, citizen groups, and non-profit Organization) and the laboratory of aquatic and marine environmental education

of Tokyo University of Marine Science and Technology established the SANRIKU ESD HRA (Manabe & Sasaki, 2009). In order to enhance aquatic marine environmental literacy, the HRA has supplied watershed activities, such as river floating, salmon hatching, salmon releasing, and smelt fish fishing for local elementary school students. In addition, the HRA has developed a variety of aquatic marine environmental education activities with the local community, such as cultivation of aquatic environmental educators, kelp learning program, and a discussion program for the future of local area.

4. **IWATE Sanriku Eco Festa:** On February 11, 2011, the IWATE Sanriku Eco Festa was held in Kamaishi City, Iwate Prefecture. In this forum, practitioners of environmental education working in the coastal area of the Iwate Prefecture met together for the first time. In the case presentation, practitioners introduced their activities and discussed their future plan. In the panel discussion, they acknowledged that wide range cooperation is necessary to promote environmental education in the abundant natural environment of Sanriku area.

Phase 2: Interruption of Aquatic Marine Environmental Education by the Great Earthquake

However, one month after holding this forum, on March 11, 2011, the Great East Japan Earthquake occurred and the practitioners received a great shock. Since AMEE was mainly carried out near rivers or the ocean, not only the practitioners, but also the affected children and parents needed time to recover from the emotional impact. Although the practitioners also suffered heavy damage in the earthquake and had to stay in evacuation shelters, they fully engaged in supporting other survivors with relief supplies, hot meals, and other assistance.

Such support activities were not directly related to environmental education, but they established the social network for affected individuals and people outside the affected area, leading to revitalization and expansion of education activities.

Phase 3: Restart of Aquatic Marine Environmental Education Six Months After the Earthquake

1. **SANRIKU Eco Vision Forum:** Under these circumstances, the SANRIKU Eco Vision Forum was held in Tono City, Iwate Prefecture, on September 4, 2011. Practitioners from the affected areas and researchers in the field of regional revitalization gathered to discuss the approach for the reconstruction

of the disaster area and their personal experience about the earthquake. They reached a consensus that the most important agenda for the Sanriku region, which experienced the tsunami every 50 years over the past 400 years, is building community capability through a wide range of collaboration. They also believed that the environmental education program around the rivers and the ocean is more necessary than ever before.

Community capability is the people's ability to observe the aquatic marine environment of the watershed, share their observation with peers, deepen understanding of the aquatic marine environment, and utilize regional resources (e.g., nature, landscape, fishery organisms of the watershed), which can be developed sustainably, spontaneously, and creatively (Sasaki, 2011b). Mutual respect and assistance between peoples can be encouraged. Watershed residents who hold universal values have the energy to achieve their goals and the ability to deal with crisis situations. These characteristics will lead to the construction of the watershed community capability (i.e., community resilience) (Sasaki, Sakana-Kun, Kawana, Osakabe & Miura 2015).

2. **Restart of River Floating Activities and Aquatic Organism Survey:** On September 11, 2011, the HRA resumed river floating activities and aquatic organism survey for the first time since the earthquake. This kind of activity was aimed at helping participants master the skill and experience for floating the river and getting closer to nature by observing the aquatic organisms. Participants were 100 in total; most of them were primary school students and their parents, including the affected families. In response to proposals from the parents of the victims of the disaster, that is "I would like to participate more in such activities" and "I want you to further enhance such practical activities", all the staff confirmed the necessity of AMEE.

Phase 4: Development of Aquatic Marine Environmental Education Aiming at Endogenous Reconstruction

1. **The Holding of the World Cherry Salmon Summit:** AMEE activities were resumed. With the enlargement of activities range and the expansion of public participation, the non-profit organization in the city of Morioka (near the city of Miyako) decided to collaborate. As a result, they consolidated the following opinions: The area affected by the tsunami is blessed with the natural environment of the forest-river-ocean nexus, where water of many clean rivers originates in mountains (e.g., the Kitakami mountain) and flows into the sea; the

implementation of FRONE enables the endogenous development of watersheds (including the headwater, estuary, and coastal areas), which is essential to the endogenous reconstruction of the coastal disaster area. According to the participants' final opinion, from January to April 2013 a series of workshops were held in the cities of Morioka and Miyako to set up an effective education program. The participants proposed that the new education program should be themed with cherry salmon, because its well-being cannot be separated from the forest-river-ocean nexus. The First World Cherry Salmon Summit was held in the city of Miyako in May 2013, to popularize the importance of the forest-river-ocean nexus (Sasaki, 2016).

This arrangement for the summit includes an excursion along the headstream of Hei River, the visit to the coastal disaster area, a foretaste of cherry salmon, and the short drama that described the life history of the fish and its connection with forest-river-ocean nexus. This summit was successfully conducted with more than 100 participants. Participating students and parents said: "I can understand the importance of the headstream;" "cherry salmon is really delicious;" "I understand that the life of cherry salmon depends on the quality of forest-river-ocean nexus". According to their comment, the authors concluded that, in order to rehabilitate the affected area, public awareness about the forest-river-ocean nexus needs to be created and raised.

2. **The 4th International Pacific Marine Educators Network Conference:** In 2014, the 4th International Pacific Ocean Educator Network Conference was held. The theme of the conference was: How can marine educators contribute to support and reconstruct the disaster areas?. This conference also arranged the excursion from the headstream of the Hei River to the coastal area. Overseas marine educators evaluated these excursions that "the natural landscape of the watershed and the coexistence of the watershed residents are important for considering what a sustainable society might be." At this conference, the importance of the nexuses including people and forests, rivers, and oceans was confirmed again (Sasaki, 2015).
3. **The Hei River Cherry Salmon MANABI Project:** In this project, stakeholders, such as watershed residents, researchers, administrators, and companies, collaborated and started to conduct the development, practice, and research of the AMEE program. This program is a whole year program to learn the life history of cherry salmon, intended for elementary and junior high school students in an area which includes the city of Miyako, Kanto area of Tokyo,

Kanagawa, and Saitama. Annual activities are as follows. May: "The World Cherry Salmon Summit in IWATE"; August: "River Flow Experience" (hands-on experience in the Hei River); November: "Cherry Salmon Hatching & Wonder of Cherry Salmon" (experiencing artificial fertilization of cherry salmon and observing the hatched salmon fry); December: "Tracing the Marked Cherry Salmon"; next April: " Releasing Event of Marked Cherry Salmon". These series of programs are called the Hei River Cherry Salmon MANABI Project and are being implemented continuously every year. In the past four years, the participants, including staff members, were over 1,000.

Phase 5: Research Results and Development of FRONE

1. **Outcomes of Local Knowledge Research and Science Research:** Between 2013 and 2016, the Ministry of Education, Culture, Sports, Science and Technology's Revitalization Project the Creation of Fisheries Research and Education Center in Sanriku was established, in collaboration with Iwate university, Tokyo University of Marine Science and Technology, and Kitasato University. They implemented the Citizen Participation Monitoring for Branding of Cherry Salmon from the Hei River. In addition, the Research Project on The Research Institute for Humanity and Nature "Long-term Sustainability through Place-Based, Small-Scale Economies: Approaches from Historical Ecology" were implemented.

As the research in the Hei River watershed, an analysis of otolith of cherry salmon (by using stable isotopic ratio and trace elements analysis), an ecological survey of its life history, an interview about the local knowledge concerning nexuses which involve people on one side and forests, rivers, and ocean on the other were also conducted. The ecological scientific knowledge of cherry salmon supports the local knowledge which is derived from the relationship between local residents and cherry salmon. The combination of these two is the basic principle of watershed aquatic marine environmental literacy. Drawing upon the researches above, the authors wish to delineate a theoretical contribution they call FIV. At its core, the authors suggest FIV involves the food that sustains human life is determined by the spatiotemporal forest-river-ocean nexus.

2. **Food Intrinsic Value:** The nature experience in the watershed environment, can help the participants realize that the food that sustains their life is closely related to a healthy forest-river-ocean nexus. Kiso (2014) listed four conditions of the

healthy forest-river-ocean nexus. Firstly, there exists a forest that can absorb and retain water, maintain high levels of organic matter, such as minerals and nutrients in the soil, and enhance the water quality. Secondly, this area should have enough spawning habitat (e.g., sand, shallows, and ridges) for creatures (e-g-, fish or other animals). Thirdly, the aquatic and marine environment should be rich in organisms that build up the bottom of food web (e.g., aquatic insects, algae, zooplankton, and small fish). The fourth condition is that few obstacles, such as dams, hinder the return of migratory fish. In other words, the healthy forest-river-ocean nexus refers to the good connection among forests, rivers, and the ocean, which allows the creature to enjoy their life. This research focused on the Hei River, which has a healthy forest-river-ocean nexus. Watershed residents have conducted the environmental conservation activities to promote the harmonious coexistence of human and nature for decades.

Thus, there exists the spatiotemporal connection between local people and the healthy forest-river-ocean nexus. Speaking of spatial connection, it means that the creatures live in and around the river, estuaries can be affected by human activities, and have an impact on human health and well-being. The river area offered enough space for the aquatic animals to spawn, the fry fish to hide from predators and find food, and also provide suitable environment for the mushroom and wild vegetable to grow. The watershed of the forest-river-ocean nexus is the habitat for lots of different creatures, which can also be served as food and source of nutrition for human. In other words, the forest-river-ocean nexus and people have a spatial connection through food.

As for the temporal connection, it implies that living creatures and people are connected from the past to the present and to the future. Nowadays, watershed residents still get food as their ancestors, through hunting and gathering in the forest, rivers, and the ocean. In order to leave the abundant biological resources for the future generations, they take the responsibility for maintaining a healthy forest-river-ocean nexus. This is how people are temporally connected with the forest-river-ocean nexus through food. Drawing upon the above analysis, the FIV is defined as follows: People's lives are supported by food derived from the spatiotemporal forest–river–ocean nexus. It is part of the Aquatic Marine Environmental Literacy, which consists of scientific knowledge and traditional knowledge.

3. **Development of FRONE Aiming to Improve Recognition and Understanding of FIV:** In order to improve, the public recognition and understanding of FIV are connected from the past to the present and to the future. Nowadays, watershed

implemented activities are listed as follows: Local government-sponsored Forest-River-Ocean Nexus Experience project to deepen the interaction with the children outside the prefecture; eco touring program in the roadside station; development of edible teaching materials to learn the ecology of the cherry salmon; education program for local college students; instruction course for local companies. The result of the questionnaire survey revealed that such education activities indeed deepened the public's recognition and understanding of FIV (Mizutani & Sasaki, 2018a).

As a result, in 2016, the practice of the Hei River Cherry Salmon MANABI Project received the Gold Award of the Japan Resilience Award. The purpose of this award is to encourage, recognize, and reward outstanding contributions to the enhancement of people's watershed awareness and the creation of resilient communities in affected areas. In 2017, this project was published as an advanced case in the UNESCO-issued marine education textbook for achieving sustainable development goals. By applying the above approaches, FRONE has become well-known in Japan and overseas. Since FRONE can be carried out in all the watershed of Japan, it can help to build up the nationwide network of collaboration (Kishi, 2002), so that educators from different places can exchange experiences and learn from each other.

Development of FRONE Seen From the Adaptive Cycle

As the authors mentioned above, FRONE was divided into 5 phases, due to the interruption which the earthquake caused. The content and outcomes of each phases are summarized below. The authors analyzed the outcomes of phases 1-5 by referring to the adaptive cycle. In phase 1, experiential activities in the Hei River watershed were carried out to pass down the rich nature of the hometown to the next generation. Such activities offered an opportunity for the participants to enhance the relationship with each other and contact the nature. It was confirmed that the participants' recognition and understanding of the nexus involving human, forests, rivers, and the ocean was improved. Thus, phase 1 can be classified as "growth or exploitation" (r) and "conservation" (K).

In phase 2, the enhanced connection among the participants and their contact with nature was temporarily interrupted by the tsunami, as a result of the 2011 Great East Japan Earthquake. Hereby, phase 2 can be considered as "collapse or release" (Ω). Despite the education activities were discontinued, practitioners built a broad network while volunteering in the affected area.

In phase 3, the SANRIKU Eco Vision Forum was held and the river experiential learning was restarted. Through the previous programs, the practitioners realized the need for conducting the aquatic marine environmental education program. In order to assess the importance of the participants' recognition and understanding of the nexus involving human, forests, rivers, and the ocean that they had before the earthquake, the practitioners decided to resume educational activities. As a result, phase 3 can be seen as "reorganization" (α). The reason why the transition from "collapse or release" (Ω) to "reorganization" (α) happens efficiently in the short term is that there existed the connection between people and people, and people and nature that had been created in phase 1, that is "growth or exploitation" (r) and "conservation" (K).

In phase 4, at the World Cherry Salmon Summit, the participants realized the forest-river-ocean nexus and their need to work together to reconstruct the watershed area. Thus, the nexuses including people and people, and people and forests, rivers, and ocean were restructured. In addition, the year-round program, the Hei River Cherry Salmon MANABI project, started, and the number of participants increased to more than 1,000 in total. As to the case of South Sweden that the authors mentioned above, phase 4 also contains these two processes: Integrating the knowledge; having the same goal-setting framework to create the goal and vision. This phase can be recognized as "growth or exploitation" (r).

In phase 5, collaborative research activities among universities confirmed the importance of establishing the nexus of forest-river-ocean and people. Since the FIV can convey the spatiotemporal forest-river-ocean nexus, various related educational activities were conducted to improve the public's recognition and understanding of this concept. Based on the authors' analysis in the case of South Sweden, phase 5 contains the following two processes: Constructing a robust social network; executing the plan. Therefore, this phase could also be classified into "conservation" phase K.

By combining the above contents of each phase with the theory of focusing on exploration target, the development of FRONE was into 5 phases. It is certain that FRONE contributed to the public's reorganization and re-recognition of nexus of forest-river-ocean and people. These achievements were developed in an interrelated way. In other words, practitioners themselves acted under universal values, developed a sense of self-efficacy, and enhanced their skills, human relations network, and mutual trust. These results correspond to autonomy, competence, and relatedness of the self-determination theory (Sasaki, 2016). It is likely that these achievements would further inspire their intrinsic motivation and creativity.

Besides, in phase 5, the practitioners defined the FIV and developed various related practices to improve public recognition and understanding of the FIV.

Furihata, Ninomiya-Rim, Noguchi, and Kobori (2013) stated that, based on the adaptive cycle, the place to learn how society and ecosystems mediate is vital to the establishment of resilience in social-ecological system. Especially for Japanese environmental education, Satoyama (Japanese term for mountains where biodiversity is increasing by people's care) and Satoumi (Japanese term for the coastal area where biodiversity is increasing by people's care) are good places for the participants to study how society and ecosystems mediate, adapt, and learn from change. This is also the case for the watershed area with the forest-river-ocean nexus. Therefore, by enhancing public recognition and understanding of the FIV through FRONE, their skills, human relation networks, and mutual trust will be strengthened in the watershed area, providing the foundation for building a resilient and sustainable society.

SOLUTIONS AND RECOMMENDATIONS

As the authors presented in the background section, public's awareness towards the coexistence of nature and human beings has been seriously diluted. This is because most of the people in developed countries have less chance to have contact with nature. They have no idea about how their consumer choices can help mitigate the negative environmental impacts the existing food system causes. Therefore, the FRONE was developed to increase public awareness of environment. However, due to the 2011 Great Earthquake, the endogenous reconstruction of the affected community became the first priority of the education program. In order to improve the community's resilience, the related improvement was made to the program. The phases 1-5 of the FRONE offered several good solutions for environmental educators or policy makers to design the system of environmental education to encourage bottom-up social action. The related recommendations are detailed in the following.

Firstly, the education program should provide various educational activities to attract the participants of different educational levels. For instance, high school and university students are likely interested in fish ecological survey, because they are able to obtain their academic achievement by carrying out field research. For preschool, kindergarten, and primary students and their parents, activities such as nature excursions, river floating, salmon hatching and releasing, fishing, tasting the wild cherry salmon and learning its life history can not only allow them to enjoy nature and have fun, but also make them realize their connection with nature, the importance of reducing the environmental impact, and the feeling of effective participation in the community development.

Secondly, outdoor environmental education should be encouraged and expanded. Ballantyne and Packer (2002) proved that combining observation with instruction is a powerful teaching strategy, especially when this allows students to understand the impact of human action on wildlife and natural inhabitant. Compared with indoor education, outdoor experiences can offer the opportunity to promote active learning and deeply impress the participants (Sasaki & Kouta, 2015). Besides, due to the distance between the classroom and the nature environment, urban students may find it difficult to understand the importance of nature to their life and the link between the environment where their food grows and their health. Thus, through outdoor environmental education, participants will improve their understanding about the interaction between human activity and the environment.

Thirdly, a network among university, local residents, cooperatives in fisheries, aquatic processing industry, retailers, non-profit organizations, and other parties should be established. Indeed, this will help them to better communicate with each other, work together, and reconstruct the community. In fact, no one is able to understand the current situation thoroughly and figure out the best solution independently. By communicating with other people from different fields, practitioners and participants can learn much about the current situation and improve their education program accordingly, as well as offer suggestions for policy makers. Combing bottom-up and top-down approaches, the local area will speed up the sustainable development of the community.

FUTURE RESEARCH DIRECTIONS

The following aspects of this paper need improvement:

1. Due to limited conditions, this education program only focused on the watershed residents in the Hei River area, which may not meet the need for good generalization. Hence, if conditions allow, a further program will be open to the residents outside the river area, and expanded to other regions and countries.
2. In order to quantify the effect of the public's enhanced recognition and understanding of the FIV on the establishment of a resilient and sustainable society, further investigation and research are needed.
3. The status of health of the aquatic ecosystem is determined by the amount of forest, the biodiversity of the river, and the discharge of sewage. Since certain aquatic species are not land-locked, the health of rivers and the ocean will

impact on the population of the species. Thus, in future studies, pollutants such as heavy metals and microplastics in the captured fish can served as an index in quantifying the status of health of the aquatic ecosystem.

4. The application of otolith tags allows to obtain the survival rate of the released cherry salmon, which can offer the basis for the bioeconomic model of the fish. Thus, such model can be used to estimate the maximum sustainable yield and associated fishing effort, and to implement the allowable catch management for this single species in Japan.

CONCLUSION

This study examined the possibility of constructing a resilient sustainable society through forest-river-ocean nexus-based education for community development (FRONE), and assessed the resilience by applying the adaptive cycle. With the application of the FET, this research divided the development of FRONE into 5 phases. The contents and achievements of each phase were also presented. In light of the adaptive cycle, it is certain that FRONE contributed to the public's reorganization and re-recognition of the nexus of forest-river-ocean and people. The achievements were developed in an interrelated way. This corresponded to autonomy, competence, and relatedness of the self-determination theory. Furthermore, in the fifth phase, the FIV was defined. Participants raised the recognition and understanding of the FIV, and increased their consciousness of spatiotemporal forest-river-ocean nexus. Therefore, their skills, human relation networks, and mutual and interactive trust was strengthened, providing the foundation for building a resilient sustainable society. These results will greatly contribute to the future development of biological economics aiming for the development of a sustainable society. The following aspects of this paper still need further improvement: (i) Due to limited conditions, this education program only focused on the watershed residents in the Hei River area, which may not meet the need of good generalization. Hence, if conditions allow, further program will be open to residents outside the river area and expanded to other regions and countries; (ii) in order to quantify the effect of the public's enhanced recognition and understanding of the FIV on the establishment of resilient sustainable society, further investigation and research are needed.

NOTE

This chapter was created based on the article (Mizutani & Sasaki, 2018b) of Japanese Journal of Environmental Education No.70.

ACKNOWLEDGMENT

Foremost, we would like to show my deepest gratitude to SANRIKU ESD Hei River Academy, Hei River fishery cooperative association, Miyako fishery commercial cooperative association, Miyako fishery cooperative association and research project of RIHN. They have provided us with valuable advice in each stage of the writing of this chapter. Especially, discussion with research members of RIHN gave us fruitful outcomes of this research. This work was financially supported by the River Foundation (No. 251215011), MEXT Revitalization Project the Creation of Fisheries Research and Education Center in Sanriku, the Nissei Foundation (wisdom and ingenuity in Mountain, River, Ocean: Practice of environmental education), Research Project on The Research Institute for Humanity and Nature "Long-term Sustainability through Place-Based, Small-Scale Economies: Approaches from Historical Ecology (No. 14200084) ", and the JSPS KAKENHI Grant Number 15K11951.

REFERENCES

Association for Resilience Japan. (2016, March 16). *Japan Resilience Award 2016*. Retrieved from http://www.resilience-jp.biz/wp-content/uploads/2016/03/RSrelease_award2016.pdf

Ballantyne, R., & Packer, J. (2002). Nature-based excursions: School students' perceptions of learning in natural environments. *International Research in Geographical and Environmental Education*, *11*(3), 218–236. doi:10.1080/10382040208667488

FAO. (2016). *How sustainability is addressed in official bioeconomy strategies at international, and regional revels an overview*. Retrieved from http://www.fao.org/3/a-i5998e.pdf

Furihata, S., Ninomiya-Rim, S., Noguchi, F., & Kobori, S. (2013). The prospective applications of resilience research and the renewal of environmental education: The power of community confronting disasters. *Japanese Journal of Environmental Education*, *22*(2), 47–58. doi:10.5647/jsoee.22.2_47

Gunderson, L. H., & Holling, C. S. (Eds.). (2002). *Panarchy: Understanding Transformations in Systems of Humans and Nature*. Washington, DC: Island Press.

Habu, J. (2018). Traditional Ecological Knowledge, Scientific Knowledge and the Concept of Resilience: Examination of Long-term Change in Landscape and Cultural Practice. In *Weaving the Knowledge of Mountains, Rivers and the Ocean: Traditional Ecological Knowledge and Ecoliteracy in Tohoku, Northern Japan* (pp. 3–12). Kanagawa: Tokai University Press.

Hernández, J. G. V., Pallagst, K., & Hammer, P. (2017). Bio economy's institutional and policy framework for the sustainable development of nature ecosystems. *Economia Cyuntural. Revista de Temas de Coyunture y Perspectivas*, 2(3), 51–104.

Hirata, O. (2013). What is true self-sustaining regeneration in the region. *Cities and Governance*, *19*, 3–8.

Holling, C. S. (1973). Resilience and stability of ecological systems. *Annual Review of Ecology and Systematics*, *4*(1), 1–23. doi:10.1146/annurev.es.04.110173.000245

Kishi, Y. (2002). What is a watershed. In Y. Konohira (Ed.), *Conservation of Watershed Environment* (pp. 70–77). Tokyo: Asakura Shoten.

Kiso, K. (2014). *Fish with two faces Cherry Salmon: Salmon remaining in the river or trout falling in the sea? Try on that mystery!* Tokyo: Seizando Shoten.

Manabe, T., & Sasaki, T. (2009). Development and practice of citizen participatory experience learning materials and its effects in the Hei River. *Memoirs of Study in Aquatic and Marine Environmental Education*, *2*(1), 41–83.

Ministry of Internal Affairs and Communications. (2015). *Fiscal yearly census population basic calculation result*. Retrieved from http://www.stat.go.jp/data/kokusei/2015/kekka.htm

Ministry of the Environment. (2016). *To connect and support Mori-Sato-Kawa-Umi (Recommendation)*. Retrieved from http://www.env.go.jp/nature/morisatokawaumi/pdf/c/teigen02.pdf

Mizutani, S., & Sasaki, T. (2018). Empirical research for environmental study materials deepened understanding relationship between forest-river-ocean and human in school setting subject "exploration" class. *Journal of Society of Clinical Education for Individual School Subjects*, *17*, 95–102.

Mizutani, S., & Sasaki, T. (2018b). Foundational Research of Education for Community Development to Emphasize the Forest-River-Ocean Nexus: Examination using an Adaptive Cycle Model to Evaluate Resilience. *Japanese Journal of Environmental Education*, *28*(2), 1–8.

Ogaki, K., & Saio, N. (2014). Tendency and problem on the reconstruction of the fishery villages in the tsunami-stricken of the Great East earthquake transregional and relative study focusing on situation before and after the earthquake and reconstruction activities. *Noson Keikaku Gakkaishi*, *33*, 197–202. doi:10.2750/arp.33.197

Olsson, P., Folke, C., & Hahn, T. (2004). Social-ecological transformation for ecosystem management: The development of adaptive co-management of a wetland landscape in southern Sweden. *Ecology and Society*, *9*(4), 2. doi:10.5751/ES-00683-090402

Ooe, T. (2014). Regional forces seen from the affected areas in 3.11. In *Proceedings of the Regional force symposium on 3.11 East Japan great earthquake disaster and intrinsic reconstruction: Consider the connection between rural areas and cities* (vol. 1, pp. 1-15). Tokyo: CSO Network.

Research Institute for Humanity and Nature. (2015). *Wisdom and ingenuity to live in mountains, rivers, and the ocean: Practice of environmental education utilizing local knowledge in the Hei River watershed, Iwate prefecture*. Retrieved from http://www.chikyu.ac.jp/fooddiversity/NISSAY/

Resilience Alliance. (Ed.). (2009). *Evaluation and management of resilience in social and ecological systems*. Kyoto: Research Institute for Humanity and Nature Resilience Project.

Sasaki, T. (2006a). Materialization of anadromous pond smelt *hypomesus nipponensis* for environmental education and outdoor ecological research: Ten years engagement with high school students. In T. Saruwatari (Ed.), *Introduction to fish environmental ecology* (pp. 262–290). Kanagawa: Tokai University Press.

Sasaki, T. (2006b). Approach aimed at systematizing Aquatic Marine Environmental Education. *Proceedings of Seminar of Society of Clinical Education for Individual School Subjects*, *5*, 13–14.

Sasaki, T. (2011a). *Theory and practice of aquatic marine environmental education*. Tokyo: Seizando Shoten.

Sasaki, T. (2011b). Let's observe, think, and act with everyone, from the SANRIKU ESD Hei River Academy: For children's tomorrow. *Proceedings of SANRIKU Eco Vision Forum Summary Collection*, *1*, 8–12.

Sasaki, T. (2015). Self-Awareness at the International Pacific Marine Educators Conference 2014 Japan. *Memoirs of Study in Aquatic and Marine Environmental Education*, *29*(2), 12-17.

Sasaki, T., Sakana-Kun, Kawana, Y., Osakabe, M., & Miura, K. (2015). Text analysis toward ocean cognition of elementary school children in devastated area, Sanriku, Japan. *Journal of Society of Clinical Education for Individual School Subjects*, *15*(1), 9–13.

Sasaki, T. (2016). Analysis of dialog processes at a development meeting for an environmental education program to emphasize the forest-river-ocean relationship in tsunami disaster areas. *Japanese Journal of Environmental Education*, *26*(1), 15–24. doi:10.5647/jsoee.26.1_15

Sasaki, T., & Kouta, A. (2015). Meaning of Cooperative Outdoor Experiential Learning in Integrated Study of Junior High School. *Journal of Society of Clinical Education for Individual School Subjects*, *15*(2), 41–47.

UNESCO. (2017). *Ocean literacy for all*. Retrieved from http://unesdoc.unesco.org/images/0026/002607/260721E.pdf

United Nations. (1992). Sustainable development. *Agenda 21*. Retrieved from https://sustainabledevelopment.un.org/outcomedocuments/agenda21

Wakana, H. (2001). Development of contemporary fish apparel in Japan. *Water Resource and Environmental Research, 14*, 1–9.

Waki, M., Sasaki, T., & Oshima, Y. (2015). A Dialogue analysis of "Exploration" in the Learning Cycle: Case of collaborative outdoor activity in "the Period of Integrated Study" at a junior high school. *Journal of Society of Clinical Education for Individual School Subjects, 15*(2), 89–98.

Walker, B., Holling, C. S., Carpenter, S. R., & Kinzig, A. (2004). Resilience, Adaptability and Transformability in Social-Ecological Systems. *Ecology and Society, 9*(2), 5. doi:10.5751/ES-00650-090205

Yamaguchi, Y. (2009). Forecast Population and Newspaper Report by Municipality in the Tohoku Region. *Quarterly Journal of Geography, 61*(3), 234–238. doi:10.5190/tga.61.234

ADDITIONAL READING

Adaptive cycle (n.d.). Retrieved July 24, 2018, from Resilience Alliance website. https://www.resalliance.org/adaptive-cycle

Bioeconomy, P. A. (2017, November 6). *Five principles for a sustainable bioeconomy.* Retrieved from https://www.balticsea-region-strategy.eu/news-room/highlights-blog/item/26-five-principles-for-a-sustainable-bioeconomy

Deci, E. L., & Ryan, M. R. (1985). *Intrinsic motivation and self-determination in human behavior*. New York, NY: Springer Science & Business Media. doi:10.1007/978-1-4899-2271-7

European Commission. (2012). *What is the bioeconomy?* Retrieved from https://ec.europa.eu/research/bioeconomy/index.cfm?pg=home

Habu, J. (2015). Mechanisms of long-term culture change and human impacts on the environment: A perspective from historical ecology, with special reference to the Early and Middle Jomon periods of prehistoric Japan. *Quaternary Research, 54*(5), 299–310. doi:10.4116/jaqua.54.299

Habu, J., Sasaki, T., & Fukunaga, M. (Eds.). (2018). *Weaving the Knowledge of Mountains, Rivers and the Ocean: Traditional Ecological Knowledge and Ecoliteracy in Tohoku, Northern Japan*. Kanagawa: Tokai University Press.

Kai, M. A. C., Balvanera, P., Benessaiah, K., Chapman, M., Diaz, S., Gómez-Baggethun, E., & Turner, N. (2016). Opinion: Why protect nature? Rethinking values and the environment. *Proceedings of the National Academy of Sciences of the United States of America*, *113*(6), 1462–1465. doi:10.1073/pnas.1525002113 PMID:26862158

Moncrieff, C. (2017). *How people of the Yukon river value salmon: A case study in the lower, middle, and upper portions of the Yukon River*. Retrieved from http://www.yukonsalmon.org/wp-content/uploads/Value-of-Salmon-Final-Report.pdf

Piccolo, J. J. (2017), Intrinsic values in nature: Objective good or simply half of an unhelpful dichotomy?, *journal of nature conservation*, 37:8-11

Resilience (n.d.). Retrieved July 24, 2018, from Resilience Alliance website. https://www.resalliance.org/resilience

Ryan, M. R., & Deci, E. L. (2000). Self-determination theory and the facilitation of intrinsic motivation, social development, and well-being. *The American Psychologist*, *55*(1), 68–78. doi:10.1037/0003-066X.55.1.68 PMID:11392867

Sasaki, T., & Miura, R. (2018). Analysis of Traditional Ecological Knowledge for Development of Aquatic Marine Environmental Education Program Aiming for Endogenous Watershed Community Development. *Noson Keikaku Gakkaishi*, *36*(4), 562–567. doi:10.2750/arp.31.562

Schuiteman, M. (2018, January 26). *Solving a Question of sustainability by studying the ear bones of Galapagos fish*. Retrieved from https://www.darwinfoundation.org/en/blog-articles/338-solving-a-question-of-sustainability-by-studying-the-ear-bones-of-galapagos-fish

Smith, A. (1894). *Munera pulveris; six essays on the elements of political economy*. New York, NY: Greenwood Press.

Suzuki, T. (1998). *The birth of community development education*. Hokkaido: Hokkaido University Press.

KEY TERMS AND DEFINITIONS

Adaptive Cycle: A fundamental unit for understanding complex systems of resilience from cells to ecosystems and societies.

Aquatic Marine Environmental Education (AMEE): An education program of aquatic marine environment based on the theory of the learning cycle. As defined in the AMEE, an aquatic environmentally literate person can observe his or her surrounding aquatic marine environment in a scientific way, inquire the related environmental problems and the well-being of human communities, grasp the required comprehensive knowledge such as aquatic marine environmental literacy, make responsible decisions and take responsible actions based on his or her outdoor learning experiences and the comprehensive knowledge, and has the capacity to convey them to as many people as possible effectively.

Food Intrinsic Value (FIV): Recognition that people live in the spatiotemporal connection of forests, rivers, and the ocean through food.

Great East Japan Earthquake: Disaster caused by the Tohoku Region Pacific Coast Earthquake that occurred on March 11, 2011.

Self-Determination Theory (SDT): A theory of motivation. It is concerned with supporting our natural or intrinsic tendencies to behave in effective and healthy ways.

Glossary

Bio-Based Products: Products wholly or partly derived from biomass, including plants, trees, and animals.

Biocombustibles: It is a mixture of organic substances that is used as fuel in internal combustion engines. Derived from biomass, organic matter originated in a biological process, spontaneous, or provoked, usable as a source of energy.

Biodiesel: Biodiesel is a liquid biofuel that is obtained from lipids of natural origin, such is the case of vegetable oils or animal fats that have had or not previous use, subjected to the transesterification process.

Bio-Ecology: It is the stream of biology which studies the connection between various living organisms and their environment.

Bioeconomic Valuation: Set of economic activities that obtain products and services, generating economic value, using biological resources as a fundamental element with a sustainable approach.

Bio-Refinery: It represents all economic actions performed from research and scientific actions focused on biotechnology.

Bio-Resources: These are biogenic resources other than fossil fuels those can be utilized for various purposes.

Bio-Technology: The utilization of biological processes for commercial and industrial purposes.

Blue Economy: A sustainable use of oceans for economic growth.

Center of Warsaw: The central borough of the city of Warsaw.

Circular Economy: Economics that refers to taking production as a basis for conversion and recycling instead of use and destruction.

Cities: The main activity is industry, commerce, and agricultural exploitation.

Company: A company is an enterprise or business that has been organized as a business and commercial society including capital and labor and whose main purpose is to obtain an economic benefit.

Construction of Beautiful Village: One of the most important Chinese policies is based on "Beautiful China" policy. Since the 18 th National Congress of the Communist Party of China, the target is development of industry and the improvement of living environment and rural culture base on the background of construction of ecological civilization.

DWG File: One of file formats used for storing two- and three-dimensional design data and metadata. It is the native format for several CAD packages including DraftSight, AutoCAD, IntelliCAD, and so on.

Eco City: Concepts of city that need to save water and energy resources and optimize its usage and management method is supposed to minimize the emission of carbon dioxide in production of energy from renewable sources.

Entrepreneurship: It is the ability of people to create new businesses. It is the person who knows how to discover, identify a specific business opportunity and then will arrange or get the necessary resources to start it and then take it to fruition.

Food Security: The condition in which people always have physical, social, and economic access to enough safe and nutritious food. This food meets dietary needs and food preferences for an active and healthy life.

Formal Institutions: They are the institutions related to constitutions, codes, laws, contracts, and other legal elements.

Green City: Comprises pro-environmental solutions, optimization and balancing of energy consumption, minimization of environmental interference and integration of greenery with buildings and urban space.

Green Economy: A resilient economy that provides a better quality of life for all within the ecological limits of the planet.

Green Urban Network: System of parks, green open spaces and public squares in the area of the city.

Green Ventilating Wedges: Green open spaces that allow the flow of fresh air.

Greening Strategy: A strategy for improving the health of residents; the investments planned for implementation until 2020 set the directions for improving the living conditions in Warsaw.

Guanzhong: One of the most important Chinese cultural birthplaces corresponding to the lower valley of the Wei River.

Informal Institutions: Are extensions, interpretations and modifications of formal rules, rules of behavior, agreements, codes of conduct or conventions and all those aspects that are related to culture.

Insecurity: Absence of security that an individual or a social group perceives regarding their image, their physical and/or mental integrity and in their relationship with the world.

Institution: Institutions are constraints that arise from human inventiveness to limit political, economic and social interactions. They include informal restrictions, such as sanctions, taboos, customs, traditions, and “codes of conduct,” as well as formal rules (constitutions, laws, property rights).

Knowledge-Based Bioeconomy: The process of transforming life science knowledge into new, sustainable, eco-efficient, and competitive products.

México: A country located in North America.

Natura 2000: Areas belonging to the network of protected natural areas in the territory of the European Union, in order to preserve specific habitats and valuable species of wildlife.

Quintuple Helix: Model that supports economic and financial aspects of the environmental challenges inherent in the causes and effects of climate change that require political and social measures that are implemented in coordination with the government, businessmen, universities, and society in favor of the environment.

Renewable Biomass: Energy sources derived from organic matter, include crop, waste, soybean, and garbage.

Smart City: Is a city in which the integration of human and technological capital takes place in order to improve the quality of life and to stimulate economic growth based on optimization in terms of natural resources management and participatory management.

Society Exclusion: It refers to the processes and situations that impede the satisfaction of the basic needs of people (work, housing, education, access to healthcare) and their participation in society.

Stress: State of mental fatigue caused by the demand for a much higher than normal performance; it usually causes various physical and mental disorders.

Sustainable Development: Inevitable and necessary changes and transformations for optimal and effective set of actions applied to improve the urban environment in connection with the development of social and economic structures, with the use of civilization achievements.

Traffic Generators: Places which attract the largest number of space users; attractive spaces that encourage people to spend time in them and use all facilities, innovations, educational elements, opportunities for recreation and practicing various forms of sport.

Urban Agriculture: Building development created to host the production of plants for the needs of the local community.

Urban Green Areas: They are urban spaces or of the periphery predominantly occupied with trees, bushes or plants, that can have different uses, either to fulfill functions of recreation, recreation, ecological, ornamentation, protection, recovery, and rehabilitation of the environment.

Urbanism: Study of the planning and management of cities and territory as well as the planning or sustainable design of a population.

Warsaw: Capital of Poland, among 22 cities in other countries, is already perceived as the world's metropolis.

Compilation of References

Agarwal, A. K. (2007). Biofuels (Alcohols And Biodiesel) Applications As Fuels For Internal Combustión Engines. *Progress in Energy and Combustion Science, 33*(3), 233–271. doi:10.1016/j.pecs.2006.08.003

Aguilar, A., Wohlgemuth, R., & Twardowski, T. (2018). Perspectives on bioeconomy. *New Biotechnology, 40*(Pt A), 181.

Aguilar, A., Magnien, E., & Thomas, D. (2013). Thirty years of European biotechnology programmes: From biomolecular engineering to the bioeconomy. *New Biotechnology, 30*(5), 410–425. doi:10.1016/j.nbt.2012.11.014 PMID:23195849

Aikawa, M., Hiraki, T., & Eiho, J. (2006). Vertical atmospheric structure estimated by heat island intensity and temporal variations of methane concentrations in ambient air in an urban area in Japan. *Atmospheric Environment, 40*(23), 4308–4315. doi:10.1016/j.atmosenv.2006.03.044

Albert, S. (2007). Transition to a bio-economy: A community development strategy discussion. *Journal of Rural and Community Development, 2*, 64–83.

Albrecht, S., Gottschick, M., Schorling, M., & Stint, S. (2012). Bio-economy at a crossroads. Way forward to sustainable production and consumption or industrialization of biomass? *GAIA Ecol. Perspect. Sci. Soc., 21*, 33–37.

Aldrich, H., & Auster, E. R. (1986). Even Dwarfs Started Small: Liabilities Of Age And Size And Their Strategic Implications. *Research in Organizational Behavior, 8*, 165–198.

Anaya, C. M. (2008). *Análisis espacial de los espacios abiertos recreativos, de uso público en el área metropolitana de Guadalajara*. Jalisco, México: Universidad de Guadalajara.

Arai, T. (1996). Changes in the hydrological environment in Tokyo. *The Journal of Geography, 105*(4), 459–474. doi:10.5026/jgeography.105.4_459

Arujanan, M., & Singaram, M. (2018). The biotechnology and bioeconomy landscape in Malaysia. *New Biotechnology, 40*(Pt A), 52-59.

Association for Resilience Japan. (2016, March 16). *Japan Resilience Award 2016*. Retrieved from http://www.resilience-jp.biz/wp-content/uploads/2016/03/RSrelease_award2016.pdf

Bahadur, G., & Morrison, M. (2010). Patenting human pluripotent cells: Balancing commercial, academic and ethical interests. *Human Reproduction (Oxford, England), 25*(1), 14–21. doi:10.1093/humrep/dep369 PMID:19897856

Ballantyne, R., & Packer, J. (2002). Nature-based excursions: School students' perceptions of learning in natural environments. *International Research in Geographical and Environmental Education, 11*(3), 218–236. doi:10.1080/10382040208667488

Basu, K. (2013). *Más allá de la mano invisible*. Fondo de Cultura Económica.

Baumol, W. J. (1990). Entrepreneurship: Productive, Unproductive, And Destructive. *Journal of Political Economy, 98*(5), 893–921. doi:10.1086/261712

Bautista, E. G. (2015). *La importancia der la Vinculación Universidades-Empresa-Gobierno en México. Volumen 5, Número 10, Revista Iberoamericana para la Investigación y el Desarrollo Económico*. RIDE.

Bazua, F. Y., & Valenti, G. (1993). Hacia Un Enfoque Amplio De Política Pública. *Revista de Administração Pública*, 84.

BECOTEPS (Bio-Economy Technology Platforms). The European Bioeconomy in 2030. Delivering Sustainable Growth by Addressing the Grand Societal Challenges. (2011). BECOTEPS. Available online: http://www.epsoweb.org/file/560

Belgian Presidency. (2010). *The Knowledge-Based Bio-Economy (KBBE) in Europe: Achievements and Challenges*. Belgian Presidency (BP). Available online: http://www.mercadosbiotecnologicos.com/ documents/the_knowledge_based_bioeconomy_kbbe_in_europe.pdf

Bioökonomierat (German Bio-economy Research and Technology Council). (2009). *Combine Disciplines, Improve Parameters, Seek out International Partnerships. First Recommendations for Research into the Bio-Economy in Germany*. Forschungs- und Technologierat Bioökonomie. Available online: http://biooekonomierat.de/fileadmin/Publikationen/Englisch/BOER_recommandation01.pdf

Bioökonomierat (German Bio-Economy Research and Technology Council). (2011). *Bio-Economy Innovation. Bio-Economy Council Report 2010.* Forschungs- und Technologierat Bioökonomie. Available online: http://biooekonomierat.de/fileadmin/Publikationen/Englisch/bioeconomy_council_ report_2010.pdf

Birch, K. (2009). The knowledge—Space dynamic in the UK bioeconomy. *Area*, *41*(3), 273–284. doi:10.1111/j.1475-4762.2008.00864.x

Birch, K. (2012). Knowledge, place, and power: Geographies of value in the bioeconomy. *New Genetics & Society*, *31*(2), 183–201. doi:10.1080/14636778.2012.662051

Biswas, P. V. (2009). December. Harnessing raindrop energy in Bangladesh. *Proceedings of the International conference on mechanical engineering*.

BMBF (German Federal Ministry for Education and Research). (2010). *Nationale Forschungsstrategie Bioökonomie 2030 (National Research Strategy Bioeconomy 2030)*. Unser Weg zu einer biobasierten Wirtschaft. Available online: https://www.bmbf.de/pub/Nationale_Forschungsstrategie_ Biooekonomie_2030.pdf

BMEL (German Federal Ministry for Food and Agriculture). (2014). *National Policy Strategy on Bioeconomy*. BMEL. Available online: http://www.bmel.de/SharedDocs/Downloads/EN/Publications/ NatPolicyStrategyBioeconomy.pdf?blob=publicationFile

Boehlje, M., & Bröring, S. (2011). The increasing multifunctionality of agricultural raw materials: Three dilemmas for innovation and adoption. *The International Food and Agribusiness Management Review*, *14*, 1–16.

Bornstein, R. D. (1968). Observations of the urban heat island effect in New York City. *Journal of Applied Meteorology and Climatology*, *7*(4), 575–582. doi:10.1175/1520-0450(1968)007<0575:OOTUHI>2.0.CO;2

Brown, N., Rappert, B., & Webster, A. (Eds.). (2012). Contested Futures. A Sociology of Prospective Techno-Science. Ashgate.

Bureau of Environment. (2016). *Tōkyō-to kankyō hakusho 2016* [White paper on the environment of Tokyo 2016]. Retrieved from http://www.kankyo.metro.tokyo.jp/basic/plan/white_paper/2016.html

Butler, T. (2015). *Overdevelopment, overpopulation, overshoot*. New York, NY: ORO Editions/Goff-Bools.

Capgemini Consulting (CC). (2010). *Roadmap zur Errichtung einer Knowledge-Based Bio-Economy. Nordrhein-Westfalen auf dem Weg in die Umsetzung*. Ministerium für Innovation, Wissenschaft und Forschung des Landes Nordrhein-Westfalen. Available online: http://www.wissenschaft.nrw.de/ fileadmin/Medien/Dokumente/Forschung/Fortschritt/Biooekonomie-Studie.pdf

Cheali, P. (2015). Upgrading of lignocellulosic biorefinery to value added chemicals: Sustainability and economics of bioethanol-derivatives. In Biomass & Bioenergy (Vol. 75, pp. 282–300). Amsterdam: Elsevier.

Chen, H. D., & Gottweis, H. (2013). Stem cell treatments in China: Rethinking the patient role in the global bio-economy. *Bioethics*, *27*(4), 194–207. doi:10.1111/j.1467-8519.2011.01929.x PMID:22092539

Chiyoda City. (2010). *Chiyoda-ku midori no jittai chōsa oyobi netsu bunpu chōsa gyōmu hōkokusho* [Report on green space and distribution of heat issues in Chiyoda Ward]. Retrieved from https://www.city.chiyoda.lg.jp/koho/machizukuri/kankyo/ryokuka/heat-island.html

Chmielewski, J. M. (2010). *Teoria urbanistyki w projektowaniu i planowaniu miast*. Warszawa: Oficyna Wydawnicza PW.

Cohen, B. (2013). *Smart City Wheel*. Retrieved from https://www.smart-circle.org/smartcity/blog/boyd-cohen-the-smart-city-wheel/

Cohen, B. (2015). *The 3 Generations Of Smart Cities. Inside the development of the technology driven city*. Retrieved from https://www.fastcompany.com/3047795/the-3-generations-of-smart-cities

Comisión Reguladora De Energía. (2015). *Desarrollo Del Marco Regulatorio Mexicano En Materia Energética*. Author.

Constitución Política De Los Estados Unidos Mexicanos, Diario Oficial De La Federación. (2017). *Constitución Política De Los Estados Unidos Mexicanos*. Author.

Contreras Comeche, R., & González García, N. (2010). La Medición Del Valor Social Y El Impacto De Los Emprendedores Sociales. In Emprendimiento, Economía Social Y Empleo. IUDESCOOP, Instituto Universitario De Economía Social Y Cooperativa Dela Universidad De Valencia.

Cooke, P. (2007). *Growth Cultures: The Global Bioeconomy and Its Bioregions*. Abingdon, UK: Routledge.

Cooke, P. (2009). The economic geography of knowledge flow hierarchies among internationally networked medical bioclusters: A scientometric analysis. *Tijdschrift voor Economische en Sociale Geografie*, *100*(3), 332–347. doi:10.1111/j.1467-9663.2009.00506.x

Dos Anjos, N. D. F. R. (1998). Source book of alternative technologies for freshwater augmentation in Latin America and the Caribbean. *International Journal of Water Resources Development*, *14*(3), 365–398. doi:10.1080/07900629849277

Echizawa, A. (2014). 1964-nen Tōkyō orinpikku to toshi keikaku [1964 Tokyo Olympics and city planning]. *Urban Housing Sciences*, *87*, 24–28.

Efken, J., Dirksmeyer, W., Kreins, P., & Knecht, M. (2016). Measuring the importance of bioeconomy in Germany: Concept and illustration. *NJAS Wageningen Journal of Life Sciences*, *77*, 9–17. doi:10.1016/j.njas.2016.03.008

Encyklopedia Warszawy. (1994). Warszawa: Wydawnictwo Naukowe PWN.

Etzkowitz, H., & Leydes, D. (2000). *The Triple Helix. University-Industry-Governement. Relations; A Laboratory for Knowledge.* Based Economic Development Easst-Review.

European Commission (EC). (2010). *Europe 2020. A Strategy for Smart, Sustainable and Inclusive Growth*. COM.

European Commission. (2010). *Innovating for Sustainable Growth: A Bioeconomy for Europe*. European Commission. Available online: http://eur-lex.europa.eu/LexUriServ/LexUriServ.do?uri=COM:2010:2020:FIN:EN:PDF

European Commission. (2012). *Innovating for Sustainable Growth - A Bioeconomy for Europe*. Luxembourg: Publications Office of the European Union.

European Commission. (2012). *Innovating for Sustainable Growth: A Bioeconomy for Europe; COM 60 final*. Brussels, Belgium: European Commission.

European Council (EC). (2000). *Lisbon Strategy. Presidency Conclusions Lisbon European Council*. Available online: http://www.europarl.europa.eu/summits/lis1_en.htm

European Union. (2018). *Bioeconomy Policy*. Retrieved from https://ec.europa.eu/research/bioeconomy/index.cfm?pg=policy

Fanuel, M., Khan, B., Singh, N., & Singh, P. (2018). Energy Production in Smart Cities by Utilization of Kinetic Energy of Vehicles over Speed Breaker. *International Journal of Civic Engagement and Social Change*, *5*(2), 1–35. doi:10.4018/IJCESC.2018040101

FAO. (2016). *How sustainability is addressed in official bioeconomy strategies at international, and regional revels an overview*. Retrieved from http://www.fao.org/3/a-i5998e.pdf

Fitzsimmons, J. (1994). Information technology and the third industrial revolution. *The Electronic Library*, *12*(5), 295–297. doi:10.1108/eb045307

FORMAS (The Swedish Research Council for Environment, Agricultural Science and Spatial Planning). (2012). *Swedish Research and Innovation. Strategy for a Bio-Based Econom*. FORMAS. Available online: http://www.formas.se/PageFiles/5074/Strategy_Biobased_Ekonomy_hela.pdf

Foro Económico Mundial. (2014). *The Bold-Ones High-Impact Entrepreneurs Who Transform Industries*. Foro Económico Mundial.

Friar, J. H., & Meyer, M. H. (2003). Entrepreneurship and start-ups in the Boston region: Factors differ- entiating high-growth ventures from micro-ven- tures. *Small Business Economics*, *21*(2), 145–152. doi:10.1023/A:1025045828202

Fuentelsaz, L., & Gonzáles, C. (2015). El Fracaso Emprendedor A Través De Las Instituciones Y La Calidad Del Emprendimiento. *Universidad Bussiness Review.*

Fuentelsaz, L., González, C., & Maícas, J. P. (2015). (Forthcoming). ¿Ayudan Las Instituciones A Entender El Emprendimiento*? Economía Industrial.*

Furihata, S., Ninomiya-Rim, S., Noguchi, F., & Kobori, S. (2013). The prospective applications of resilience research and the renewal of environmental education: The power of community confronting disasters. *Japanese Journal of Environmental Education*, *22*(2), 47–58. doi:10.5647/jsoee.22.2_47

Furukawa, K. (2013). *Shuto kōsoku dōro no saisei* [Renewal of Metropolitan Expressway]. Reference.

German Presidency (GP). (2007). *En Route to the Knowledge-Based Bio-Economy*. Cologne Paper. Available online https://dechema.de/dechema_media/Cologne_Paper-p-20000945.pdf

Grochulska-Salak M., (2017). Rolnictwo miejskie jako element krystalizujący przestrzeń suburbiów. *Kwartalnik Naukowy Uczelni Vistula, 4*.

Grochulska-Salak, M. (2011). Projektowanie urbanistyczno-architektoniczne w mieście uwzględniające zasady zrównoważonego rozwoju w Energia i Budynek 02(45)201, Warszawa 2011.

Grochulska-Salak, M., Zielonko-Jung, K., & Zinowiec-Cieplik, K. (2018). Prace Naukowe Uniwersytetu Ekonomicznego we Wrocławiu nr XXX: Gospodarka przestrzenna – stan obecny i wyzwania przyszłości. Academic Press.

Gu, J. (2016). Application of UAV Photography to Ownership Affirmation, Registration and Certification for Right to Manage Contracted Rural Land—Take Fengyang County as a Case. *Bulletin of Surveying and Mapping,* (5), 94-99.

Gunderson, L. H., & Holling, C. S. (Eds.). (2002). *Panarchy: Understanding Transformations in Systems of Humans and Nature*. Washington, DC: Island Press.

Guranowska-Gruszecka, K. (2017), *Shaping of the Centre of Warsaw –history of activities and views, visions of future.* Szczecin wyd. PAN DOI: 1021005/pif.2017.30.A-01

Guranowska-Gruszecka, K. (2013). *Śródmieście Warszawy w XX wieku, Warszawa wyd. Szkoła Wyższa im.* Bogdana Janskiego.

Gustafsson, M., Stoor, R., & Tsvetkova, A. (2011). Sustainable bio-economy: potential, challenges and opportunities in Finland. *SITRA Studies, 51*, 64.

Gzell, S. (2015). *Wykłady o współczesnej urbanistyce*. Warszawa: Oficyna Wydawnicza PW.

Hábitat & Keever. (2005). *Ciudades "Sobre indicadores de Sustentabilidad Urbana y Observatorios.* Red Nacional de Investigación Urbana.

Habu, J. (2018). Traditional Ecological Knowledge, Scientific Knowledge and the Concept of Resilience: Examination of Long-term Change in Landscape and Cultural Practice. In *Weaving the Knowledge of Mountains, Rivers and the Ocean: Traditional Ecological Knowledge and Ecoliteracy in Tohoku, Northern Japan* (pp. 3–12). Kanagawa: Tokai University Press.

Hansen, J. (2014). The Danish biofuel debate: Coupling scientific and politico-economic claims. *Sci. Cult., 23*(1), 73–97. doi:10.1080/09505431.2013.808619

Hansen, T., & Winther, L. (2011). Innovation, regional development and relations between high- and low-tech industries. *European Urban and Regional Studies*, *18*(3), 321–339. doi:10.1177/0969776411403990

Harvey, F. (2010). Green vision: the search for the ideal eco-city. *Financial Times*. Retrieved from https://www.ft.com/content/c13677ce-b062-11df-8c04-00144feabdc0

Hekkert, M., Suurs, R. A. A., Negro, S. O., Kuhlmann, S., & Smits, R. E. H. M. (2007). Functions of innovation systems: A new approach for analysing technological change. *Technological Forecasting and Social Change*, *74*(4), 413–432. doi:10.1016/j.techfore.2006.03.002

Hernández, J. G. V., Pallagst, K., & Hammer, P. (2017). Bio economy's institutional and policy framework for the sustainable development of nature ecosystems. *Economia Cyuntural. Revista de Temas de Coyunture y Perspectivas*, *2*(3), 51–104.

Hilgartner, S. (2007). *Making the bioeconomy measurable: Politics of emerging anticipatory machinery*. Academic Press.

Hirata, O. (2013). What is true self-sustaining regeneration in the region. *Cities and Governance*, *19*, 3–8.

Hitt, M. A., Ireland, R. D., Sirmon, D. G., & Trahms, C. A. (2011). Strategic Entrepreneurship: Creating Value For Individuals, Organizations, And Society. *The Academy of Management Perspectives*, *25*(2), 57–75.

Hofstede, G. (2011). *Culture's consequences* (1st ed.). Thousand Oaks, CA: Sage Publ.

Holling, C. S. (1973). Resilience and stability of ecological systems. *Annual Review of Ecology and Systematics*, *4*(1), 1–23. doi:10.1146/annurev.es.04.110173.000245

Hompo, Y., & Yagasaki, N. (2015). Japan's inbound strategies using 2020 Tokyo Olympic and Paralympic Games based on the experiences of the past games [in Japanese]. *The International Journal of Tourism Science, 8*, 3-11.

Horie, K. (1996). Planning process of expressway in Tokyo [in Japanese]. *Doboku Keikakugaku Kenkyu, Ronbunshu, 13*, 1–22. doi:10.2208/journalip.13.1

Ikeda, M. (2017). A review of nightlife literature and suggestions for future research in Japan. *Geographical Space*, *10*(2), 67–84.

International Olympic Committee. (2013). *Report of the 2020 evaluation commission*. Retrieved from https://stillmed.olympic.org/Documents/Host_city_elections/2020_Evaluation_Commission_report.pdf

Ishibashi, N., & Takahashi, T. (2015). *Background of the exclusive use of the elevated Metropolitan Expressway over Nihonbashi River.* The Institute of Civil Engineering of the Tokyo Metropolitan Government Annual Report.

Johnston, B. F., & Mellor, J. W. (1961). The role of agriculture in economic development. *The American Economic Review*, *51*(4), 566–593.

Kearnes, M. (2013). Performing synthetic worlds: Situating the bioeconomy. *Science & Public Policy*, *40*(4), 453–465. doi:10.1093cipolct052

Keegan, D., Kretschmer, B., Elbersen, B., & Panoutsou, C. (2013). Cascading use: A systematic approach to biomass beyond the energy sector. *Biofuels, Bioproducts & Biorefining*, *7*(2), 193–206. doi:10.1002/bbb.1351

Khan, B., & Singh, P. (2017). Selecting a Meta-Heuristic Technique for Smart Micro-Grid Optimization Problem: A Comprehensive Analysis. *IEEE Access: Practical Innovations, Open Solutions*, *5*, 13951–13977. doi:10.1109/ACCESS.2017.2728683

Kishi, Y. (2002). What is a watershed. In Y. Konohira (Ed.), *Conservation of Watershed Environment* (pp. 70–77). Tokyo: Asakura Shoten.

Kiso, K. (2014). *Fish with two faces Cherry Salmon: Salmon remaining in the river or trout falling in the sea? Try on that mystery!* Tokyo: Seizando Shoten.

Klein, W., Lankhuizen, R. M., & Gilsing, V. (2005). A system failure framework for innovation policy design. *Technovation*, *25*(6), 609–619. doi:10.1016/j.technovation.2003.11.002

Kleszcz, J. (2016). Farma w mieście – wizja rolnictwa XXI wieku. *ARCHITECTURAE et ARTIBUS*, *3*. Retrieved from http://www.wa.pb.edu.pl/uploads/downloads/Architektura-3-2016---artykul-6--do-internetu-.pdf

Kondo, H. (2009). City canopy [in Japanese]. *Tenki*, *56*(8), 677–678.

Lainez, M., González, J. M., Aguilar, A., & Vela, C. (2017). Spanish strategy on bioeconomy: Towards a knowledge based sustainable innovation. *New Biotechnology*. PMID:28552816

Łaskarzewska, M. (2015). Piękno i użyteczność "Polskiej Zielonej Ściany". *Zeszyty naukowe uczelni Vistula, 42*(4), 60 – 69.

Lee, J., Harashima, Y., Lee, D. K., & Morita, T. (1995). A comparative analysis of processes used in developing environmental policies in Japan and Korea. *Environmental Sciences, 8*(2), 181–192.

Levidow, L., Birch, K., & Papaioannou, T. (2013). Divergent paradigms of European agro-food innovation: The knowledge-based bio-economy (KBBE) as an R & D agenda. *Science, Technology & Human Values, 38*(1), 94–125. doi:10.1177/0162243912438143

Ley De Promoción Y Desarrollo De Los Bionergéticos. (2008). *Ley De Promoción Y Desarrollo De Los Bionergéticos*. Publicado En El Diario Oficial De La Federación El Primero De Febrero De.

Liñán, F., Fernández, J., & Romero, I. (2013). Necessity And Opportunity Entrepreneurship: The Mediating Effect Of Culture. *Revista de Economía Mundial, 33*, 21–47.

Lokko, Y., Heijde, M., Schebesta, K., Scholtès, P., Van Montagu, M., & Giacca, M. (2017). Biotechnology and the bioeconomy—Towards inclusive and sustainable industrial development. *New Biotechnology*. PMID:28663120

Lösch, A., & Schneider, C. (2016). Transforming power/knowledge apparatuses: The smart grid in the German energy transition. *Innov. Euro. J. Soc. Sci. Res., 29*(3), 262–284. doi:10.1080/13511610.2016.1154783

Low, S. A., & Isserman, A. M. (2009). Ethanol and the local economy: Industry trends, location factors, economic impacts, and risks. *Economic Development Quarterly, 23*(1), 71–88. doi:10.1177/0891242408329485

Maeda, C., Kondo, A., Akikawa, M., Kaga, A., & Inoue, Y. (2008, March): *Simulation of urban heat island in Hyogo*. Paper presented at the meeting of the Society of Heating, Air-conditioning Sanitary Engineers of Japan, Osaka, Japan.

Mahoney, J. (2012). Property rights theory. In Economic Fundations of Strategy (pp. 158–199). Champaign, IL: Academic Press.

Makki, N., & Pop-Iliev, R. (2011). Piezoelectric power generation in automotive tires. *Proceedings of the Smart Materials & Structures/NDT in Aerospace/NDT in Canada*.

Małyska, A., & Jacobi, J. (2017). Plant breeding as the cornerstone of a sustainable bioeconomy. *New Biotechnology*. PMID:28690155

Manabe, T., & Sasaki, T. (2009). Development and practice of citizen participatory experience learning materials and its effects in the Hei River. *Memoirs of Study in Aquatic and Marine Environmental Education*, *2*(1), 41–83.

Marco, O., & Edmond, P. (2016). *La ségrégation urbaine. Editions. La Découverte*. Paris: Francia.

Marsden, T. (2012). Towards a real sustainable agri-food security and food policy: Beyond the ecological fallacies? *The Political Quarterly*, *83*(1), 139–145. doi:10.1111/j.1467-923X.2012.02242.x

Martin, S., & Shrivastava, K. K. (2013). Feasibility of rainwater harvesting in high rise building for power generation. *International Journal of Engineering Trends and Technology*, *4*, 522–527.

Marulanda, F., & Morales, S. (2016). Entorno Y Motivaciones Para Emprender. *Revista Escuela De Administración De Negocios,* 8.

Maruyama, M., & Nakagawa, Y. (2004). Shutoken seibi keikaku ni okeru ryokuchi seisaku no hensen ni kansuru kisoteki kousatsu [Basic consideration about the transition of green policies in Metropolitan Region development]. In *Proceedings of the Meeting of Japan Society of Civil Engineers* (Vol. 59, pp. 4-334). Tokyo, Japan: Japan Society of Civil Engineers.

Masumoto, K. (2007). Space-time characteristics of air temperature distribution and heat island situation in Osaka City. *Journal of Urban Living and Health Association*, *51*(4), 215–224.

Mathews, J. A. (2009). From the petroeconomy to the bioeconomy: Integrating bioenergy production with agricultural demands. *Biofuels, Bioproducts & Biorefining*, *3*(6), 613–632. doi:10.1002/bbb.181

Matsuoka, R. (2017). *Toshi ni okeru ryokuchi ni kansuru seido keikaku no hensen to Tōkyō 23-ku no ryokuchi seisaku no tokuchō* [Transition of institution and planning related to urban green spaces and the features of green policy in Tokyo's 23 wards] (Unpublished graduation thesis). Tokyo, Japan: Waseda University.

McCormick, K., & Kautto, N. (2013). The bioeconomy in Europe: An overview. *Sustainability*, *5*(6), 2589–2608. doi:10.3390u5062589

Medina, I., Chavez, N. Y., & Jauregui, J. (2012). Biodiesel, Un Combustible Renovable. *Investigación Y Ciencia, 20*, 62-70.

Miki, T. (2017). Redevelopment projects in Shibuya Tokyu Corporation [in Japanese]. *Journal of JSEE, 65*(6), 41–43. doi:10.4307/jsee.65.4_1

Minato City. (2017). *Minato-ku midori no jittai chōsa* [Report of Green Investigation in Minato]. Retrieved from https://www.city.minato.tokyo.jp/ryokukasuishin/kankyo-machi/kankyo/chosa/9midorinojittai.html

Ministry of Internal Affairs and Communications. (2015). *Fiscal yearly census population basic calculation result.* Retrieved from http://www.stat.go.jp/data/kokusei/2015/kekka.htm

Ministry of Land, Infrastructure, Transport and Tourism. (2011). *Shuto-ken seibi hou ni motozuku daitoshi-ken no minaoshi* [Re-examining the Tokyo metropolitan regional policy based on Metropolitan Region Development Law]. Retrieved from http://www.mlit.go.jp/common/000139692.pdf

Ministry of Land, Infrastructure, Transport and Tourism. (2017). Zenkoku okujō hekimen ryokuka sekō jisseki chōsa no kekka hōkoku [Report of investigation of rooftop and wall greening in Japan] [Data file]. Retrieved from http://www.mlit.go.jp/report/press/toshi10_hh_000257.html

Ministry of the Environment. (2014). *2020-nen orinpikku pararinpikku Tōkyō taikai wo keiki toshita kankyō hairyo no suishin ni tsuite* [Report on promoting the environment policy for the 2020 Olympics and Paralympics]. Retrieved from https://www.env.go.jp/press/18532.html

Ministry of the Environment. (2016). *To connect and support Mori-Sato-Kawa-Umi (Recommendation).* Retrieved from http://www.env.go.jp/nature/morisatokawaumi/pdf/c/teigen02.pdf

Miranda, C. (1997). Filosofía y Medio Ambiente. Una Aproximación teórica. Taller Abierto, México, D.F.

Mizuno, M. (1993). Subjects on environment in 24-hour city: Discussion on urban heat island [in Japanese]. *The Journal of the Acoustical Society of Japan, 49*(11), 832–838.

Mizutani, S., & Sasaki, T. (2018). Empirical research for environmental study materials deepened understanding relationship between forest-river-ocean and human in school setting subject "exploration" class. *Journal of Society of Clinical Education for Individual School Subjects, 17*, 95–102.

Mizutani, S., & Sasaki, T. (2018b). Foundational Research of Education for Community Development to Emphasize the Forest-River-Ocean Nexus: Examination using an Adaptive Cycle Model to Evaluate Resilience. *Japanese Journal of Environmental Education*, *28*(2), 1–8.

Molina, C. (2012). Estudio de la composición y estabilidad de biodiesel obtenido a partir de aceites vegetales limpios y procedentes de aceites de fritura. Servicio de publicaciones de la Universidad de la Laguna.

Moreau, N. J. (2005). Public images of chemistry. *Chemistry International*, *27*(4).

More, N. N. (2013). Finite Element Analysis of Piezoelectric Cantilever. *International Journal of Innovations in Engineering and Technology*, *2*(3), 100–105.

Morrison, M., & Cornips, L. (2012). Exploring the role of dedicated online biotechnology news providers in the innovation economy. *Science, Technology & Human Values*, *37*(3), 262–285. doi:10.1177/0162243911420581

Mumtaz, Z., Bowen, S., & Mumtaz, R. (2012). Meanings of blood, bleeding and blood donations in Pakistan: Implications for national vs. global safe blood supply policies. *Health Policy and Planning*, *27*(2), 147–155. doi:10.1093/heapol/czr016 PMID:21372061

Muñoz & Carmen. (1996). Principales tendencias y Modelos de la +Educación Ambiental en el sistema escolar. Revista Iberoamericana de Educación, 11.

MWK BW (Ministry of Science, Research and Art Baden-Württemberg). (2013). *Bioökonomie im System Aufstellen. Konzept für Eine Baden-Württembergische Forschungsstrategie "Bioökonomie"* [Concept for the Implementation of a Research Strategy on the Bioeconomy in Baden-Württemberg]. MWK BW. Available online: https://mwk.baden-wuerttemberg.de/fileadmin/redaktion/m-mwk/intern/dateien/pdf/ Forschung/ Konzept_Forschungsstrategie_Biooekonomie.pdf

Nakabayashi, S., Kishī, T., & Ōsawa, M. (2012). On the rooftop gardening policy in the 23 wards of Tokyo: Focused on the actual condition of the public buildings [in Japanese]. *Journal of the City Planning Institute of Japan*, *47*(3), 475–480.

Nakamura, S., & Oki, T. (2009). The philosophy and chronicle of abolished urban river in 36'report [in Japanese]. *Proceedings of Hydraulic Engineering, JSCE*, *53*, 565–570.

North, D. (1990). *Institutions, Institutional Change and Economic Performance*. Cambridge, UK: Cambridge University Press. doi:10.1017/CBO9780511808678

OECD. (2009). *The Bioeconomy to 2030: Designing a Policy Agenda*. Paris: OECD Publishing; doi:10.1787/9789264056886-

OECD. (2017). *OECD Stats FAO Agricultura*. Retrieved from Http://Stats.Oecd.Org/Viewhtml.Aspx?Queryid=58648&Vh=0000&Vf=0&L&Il=&Lang=En#

OECD. (2018). *Meeting Policy Challenges for a Sustainable Bioeconomy*. Paris: OECD Publishing; doi:10.1787/9789264292345-

Ogaki, K., & Saio, N. (2014). Tendency and problem on the reconstruction of the fishery villages in the tsunami-stricken of the Great East earthquake transregional and relative study focusing on situation before and after the earthquake and reconstruction activities. *Noson Keikaku Gakkaishi*, *33*, 197–202. doi:10.2750/arp.33.197

Oke, T. R. (1979). Review of urban climatology 1973-1976. *WMO-Technical Note*, *169*, 1–100.

Ollikainen, M. (2014). Forestry in bioeconomy—Smart green growth for the humankind. *Scandinavian Journal of Forest Research*, *29*(4), 360–366. doi:10.1080/02827581.2014.926392

Olsson, P., Folke, C., & Hahn, T. (2004). Social-ecological transformation for ecosystem management: The development of adaptive co-management of a wetland landscape in southern Sweden. *Ecology and Society*, *9*(4), 2. doi:10.5751/ES-00683-090402

Ooe, T. (2014). Regional forces seen from the affected areas in 3.11. In *Proceedings of the Regional force symposium on 3.11 East Japan great earthquake disaster and intrinsic reconstruction: Consider the connection between rural areas and cities* (vol. 1, pp. 1-15). Tokyo: CSO Network.

OPST (Office of Science and Technology Policy). (2011). *National Bioeconomy Blueprint: Public Comment*. OPST. Available online: https://obamawhitehouse.archives.gov/administration/ eop/ostp/library/bioeconomy

Organisation for Economic Co-operation and Development (OECD). (2006). The Bioeconomy to 2030.Designing a Policy Agenda. Paris, France: OECD.

Organisation for Economic Co-operation and Development (OECD). (2009). The Bioeconomy to 2030: Designing a Policy Agenda. Paris, France: OECD.

Ozaki, M. (2002). A Study of the Process of Forming Sport Policy in Japan [in Japanese]. *Hitotsubashi University Research Series Humanities*, *39*, 159–252.

Pałac Saski w Warszawie zostanie odbudowany? (n.d.). Retrieved from http://historia.org.pl/2011/05/06/palac-saski-w-warszawie-zostanie-odbudowany/

Patel, I., Siores, E., & Shah, T. (2010). Utilisation of smart polymers and ceramic based piezoelectric materials for scavenging wasted energy. *Sensors and Actuators. A, Physical*, *159*(2), 213–218. doi:10.1016/j.sna.2010.03.022

Peng, M. (2012). *Enfatizando Las Instituciones, La Cultura Y La Ética*. Gestión Estratégica. Editoral Cengage Learning.

Pollack, A. (2012). White house promotes a bioeconomy. *N. Y. Times*.

Ponte, S. (2009). From fishery to fork: Food safety and sustainability in the 'virtual' knowledge-based bio-economy (KBBE). *Sci. Cult.*, *18*(4), 483–495. doi:10.1080/09505430902873983

Popovici, D. (2008). *Modeling and simulation of piezoelectric devices*. In *Modelling and Simulation*. InTech. doi:10.5772/5968

Pozzi, M., & Zhu, M. (2011). Plucked piezoelectric bimorphs for knee-joint energy harvesting: Modelling and experimental validation. *Smart Materials and Structures*, *20*(5), 055007. doi:10.1088/0964-1726/20/5/055007

Pülzl, H., Kleinschmit, D., & Arts, B. (2012). From a fossil-fuel to a biobased economy: The politics of industrial biotechnology. *Environment and Planning. C, Government & Policy*, *30*(2), 282–296. doi:10.1068/c10209

Raibaud, Y. (2015). *La ville faite par et pour les hommes*. Editions Belin, Paris, Francia. Retrieved from http://iitej.blogspot.mx/2012/11/cantidad-de-areas-verdes-de-guadalajara.html#more

Ramsden, Ed., & Dix, C. (2003). Low-Pass Filtering for Vibration Sensors. *Sensor Technology and Design*. Retrieved from http://archives.sensorsmag.com/articles/0203/33

Reglamento de Parques, Jardines y Recursos Forestales para el Municipio de Guadalajara. (2014). Gaceta Municipal de Guadalajara.

Rembio. (2009). *Expert Opinion Based On Delivered Questionnaire*. Author.

Research Institute for Humanity and Nature. (2015). *Wisdom and ingenuity to live in mountains, rivers, and the ocean: Practice of environmental education utilizing local knowledge in the Hei River watershed, Iwate prefecture*. Retrieved from http://www.chikyu.ac.jp/fooddiversity/NISSAY/

Resilience Alliance. (Ed.). (2009). *Evaluation and management of resilience in social and ecological systems*. Kyoto: Research Institute for Humanity and Nature Resilience Project.

Reynolds, P., Bygrave, W. D., Autio, E., Cox, L. W., & Hay, M. (2003). *Global Entrepreneurship Monitor 2002 executive report*. Wellesley, MA: Babson College.

Riegelhaupt, E., Odenthal, J., & Janeiro, L. (2016). *Diagnóstico de la situación actual del biodiésel en México y escenarios para su aprovechamiento*. Informe Final CONFIDENCIAL red Mexicana de Bioenergía. Ecofys 2016 por orden de: Banco Interamericano de Desarrollo para SENER

Roseland, M. (1997). Dimensions of the Eco-city. *Cities (London, England)*, *14*(4), 197–202. doi:10.1016/S0264-2751(97)00003-6

Rosemann, A. (2014). Standardization as situation-specific achievement: Regulatory diversity and the production of value in intercontinental collaborations in stem cell medicine. *Social Science & Medicine*, *122*, 72–80. doi:10.1016/j.socscimed.2014.10.018 PMID:25441319

Rutz, D., Thebaud, A., Janssen, R., Segura, S. A., Riegelhaupt, E., & Ballesteros, M. ... Bravo G. (2009). *Biofuel Policies And Legislation In Latin America*. WIP Renewable Energies; Report Of The EU Project Biotop (FP7); Contract No.: 213320.

Sachsenmeier, P. (2016). Industry 5.0—The Relevance and Implications of Bionics and Synthetic Biology. *Engineering*, *2*(2), 225–229. doi:10.1016/J.ENG.2016.02.015

SAGARPA. (2010). *Monografía de cultivos: palma de Aceite. Subsecretaría de Fomento a los Agronegocios*. Author.

SAGARPA. (2016). *Direccion General de Fibras Naturales y Biocombustibles. BIODIÉSEL 06052016. Presentación realizada en el Taller de Biodiésel*. Paquete Tecnológico de Higuerilla y Paquete Tecnológico de Jatropha.

SAGARPA. (2017). *Impulsa SAGARPA Producción De Biocombustibles En México*. Author.

Saito, T. (2014). Orinpikku pararinpikku to kankyō risku no kanri [Olympics, Paralympics and the management of environmental risks]. *Sompo Japan Nipponkoa Risk Management Report*, *126*, 1–11.

Sakakibara, Y. (2001). Comparison between the effect of heating from urban surface and that of mechanical mixing of urban atmosphere to heat island [in Japanese]. *Tenki*, *48*(5), 305–311.

Salgado, C. E. (2003). *Teoría de costos de transacción: una breve reseña. Cuadernos de Administración*. Retrieved from http://www.redalyc.org/articulo.oa?id=20502604

Sánchez, L. A., & Pérez, E. (2015). Las Entidades De Economía Social Como Protagonistas De Un Nuevo Modelo De Emprendimiento Y Medidas Legales De Apoyo Al Emprendimiento, CIRIEC-España. *C.I.R.I.E.C. España*, *84*, 35–62.

Sasaki, T. (2015). Self-Awareness at the International Pacific Marine Educators Conference 2014 Japan. *Memoirs of Study in Aquatic and Marine Environmental Education*, *29*(2), 12-17.

Sasaki, T. (2006a). Materialization of anadromous pond smelt *hypomesus nipponensis* for environmental education and outdoor ecological research: Ten years engagement with high school students. In T. Saruwatari (Ed.), *Introduction to fish environmental ecology* (pp. 262–290). Kanagawa: Tokai University Press.

Sasaki, T. (2006b). Approach aimed at systematizing Aquatic Marine Environmental Education. *Proceedings of Seminar of Society of Clinical Education for Individual School Subjects*, *5*, 13–14.

Sasaki, T. (2011a). *Theory and practice of aquatic marine environmental education*. Tokyo: Seizando Shoten.

Sasaki, T. (2011b). Let's observe, think, and act with everyone, from the SANRIKU ESD Hei River Academy: For children's tomorrow. *Proceedings of SANRIKU Eco Vision Forum Summary Collection*, *1*, 8–12.

Sasaki, T. (2016). Analysis of dialog processes at a development meeting for an environmental education program to emphasize the forest-river-ocean relationship in tsunami disaster areas. *Japanese Journal of Environmental Education*, *26*(1), 15–24. doi:10.5647/jsoee.26.1_15

Sasaki, T., Sakana-Kun, Kawana, Y., Osakabe, M., & Miura, K. (2015). Text analysis toward ocean cognition of elementary school children in devastated area, Sanriku, Japan. *Journal of Society of Clinical Education for Individual School Subjects*, *15*(1), 9–13.

Sasaki, T., & Kouta, A. (2015). Meaning of Cooperative Outdoor Experiential Learning in Integrated Study of Junior High School. *Journal of Society of Clinical Education for Individual School Subjects*, *15*(2), 41–47.

Sayre, R. (2013). Initial risk assessment of genetically modified (GM) algae for commodity-scale cultivation. Algal Research, 2(1), 66-77.

Scarlat, N., Dallemand, J.-F., Monforti-Ferrario, F., & Nita, V. (2015). The role of biomass and bioenergy in a future bioeconomy: Policies and facts. *Environmental Development*, *15*, 3–34. doi:10.1016/j.envdev.2015.03.006

Schaffers, H. (2011). Smart Cities and the Future Internet: Towards Cooperation Frameworks for Open Innovation. In *The The Future Internet*. Future Internet Assembly, Springer. doi:10.1007/978-3-642-20898-0_31

Schmidt, O., Padel, S., & Levidow, L. (2012). The bio-economy concept and knowledge base in a public goods and farmer perspective. *Bio-Based and Applied Economics*, *1*(1), 47–63.

Schumpeter, J. A. (1934). The Thory Of Economic Development: An Inquiry Into Profits, Capital, Credit, Interest And The Business Cycle. Harvard Economic Studies, 46.

Schütte, G. (2017). What kind of innovation policy does the bioeconomy need? *New Biotechnology*. PMID:28458016

Schwab, K. (2017). *The fourth industrial revolution*. Crown Business.

SEDATU. (2015). Available: http://www.sedatu.gob.mx/sraweb/

Shibuya City. (2004). *Shizen kankyō chōsa hōkokusho* [Report on the investigation of the environment]. Retrieved from https://www.lib.city.shibuya.tokyo.jp/?action=common_download_main&upload_id=514

Shibuya City. (2016). *Shibuya-ku midori no seibi hōshin* [Green policy in Shibuya Ward]. Retrieved from https://www.city.shibuya.tokyo.jp/kusei/shisaku/ku_keikaku/green_plan.html

Siegmeier, T., & Möller, D. (2013). Mapping research at the intersection of organic farming and bioenergy—A scientometric review. *Renewable & Sustainable Energy Reviews*, *25*, 197–204. doi:10.1016/j.rser.2013.04.025

Singh, P., & Khan, B. (2017). Smart Microgrid Energy Management Using a Novel Artificial Shark Optimization. Complexity. doi:10.1155/2017/2158926

Smolarski, N. (2012). *High-value opportunities for lignin: Unlocking its potential.* Frost & Sullivan. Retrieved from www.greenmaterials.fr/wp-content/uploads/2013/01/High-value- Opportunities-for-Lignin-Unlocking-its-Potential-Market-Insights.pdf

Śpiewakowski, E. R., & Korczyński, M. (1999). *Rozwój cywilizacji ludzkiej w kontekście sukcesji ekologicznej, „Hipoteza ekologii uniwersalistycznej", Centrum Uniwersalizmu przy Uniwersytecie Warszawskim Polska Federacja Życia.* Warszawa: Drukarnia Jerzego Kosińskiego.

Staffas, L., Gustavsson, M., & McCormick, K. (2013). Strategies and policies for the bioeconomy and bio-based economy: An analysis of official national approaches. *Sustainability*, *5*(6), 2751–2769. doi:10.3390u5062751

Stam, E., Suddle, K., Hessels, J., & Van Stel, A. J. (2009). High-Growth Entrepreneurs, Public Policies, And Economic Growth. Public Policies For Fostering Entrepreneurship: A European Perspective. International Studies In Entrepreneurship, 22, 91-110.

Stefańska & Wieczorek. (2018). *Diagnoza użytkowania Pola Mokotowskiego.* Raport z badań psychologii Środowiskowej.

Szulczewska, B., & Bruszewska, K. (2013). *Urban Agriculture in the city of Warsaw.* Poland: Academic Press.

Szwed, D., & Maciejewska, B. (2011). *Zrównoważony rozwój metropolii.* Gdańsk: Fundacja Przestrzenie Dialogu.

Takeuchi, K., & Lee, D. K. (1988). A framework for environmental management planning [in Japanese]. *Journal of the Japanese Institute of Landscape Architects*, *52*(2), 95–104. doi:10.5632/jila1934.52.95

Takeuchi, T. (2012). A study on the potential of Green Structure Plan as comprehensive spatial control plan [in Japanese]. *Journal of the Japanese Institute of Landscape Architecture*, *75*(5), 601–604. doi:10.5632/jila.75.601

Takeuchi, T., & Ishikawa, M. (2008). A study on the green space policies on the fringe of Tokyo wards area in the 1950's and 60's [in Japanese]. *Journal of the City Planning Institute of Japan*, *43*(3), 199–204.

Tapia, P. (2017). *Biodiésel, a la espera de brillar en México*. Retrieved from http://www.milenio.com/negocios/biodiesel-a-la-espera-de-brillar-en-mexico

Telba, A., & Ali, W. G. (2012), July. Modelling and simulation of piezoelectric energy harvesting. *Proceedings of the World Congress on Engineering* (*Vol. 2*, pp. 4-6). Academic Press.

Thierry, P. (2016). *Terre urbaine. Cinq défis pour le devenir urbain de la planéte. Editions. La Découverte*. Paris: Francia.

Tinaikar, A. (2013). Harvesting Energy from Rainfall. *International Journal of Renewable and Sustainable Energy*, *2*(3), 120–132.

Tokyo Metropolitan Government. (2017). *New Tokyo. New tomorrow. The action plan for 2020*. Retrieved from http://www.metro.tokyo.jp/english/about/plan/index.html

Tokyu Corporation. (2013). *Shibuya-eki minami gaiku purojekkuto ni kansuru toshi keikaku no kettei ni tsuite* [Notification of the decision about the south district project at Shibuya Station]. Retrieved from https://www.tokyu.co.jp/file/130123-3.pdf

Trzepacz, P. (Ed.). (2012). Zrównoważony rozwój – wyzwania globalne. Podręcznik dla uczestników studiów doktoranckich, Instytut Geografii i Gospodarki Przestrzennej UJ.

Tsuboi, S. (2003). A study on change of water area and water activity projects in Tokyo ward area [in Japanese]. *Comprehensive Urban Studies*, *82*, 19–34.

Tsunematsu, N. (2014). *Tōkyō ni okeru hiito airando no jittai to shonetsu taisaku ni tsuite* [Heat island in Tokyo and countermeasures against heat]. Retrieved from https://www.tokyokankyo.jp/kankyoken/wp-content/uploads/sites/3/2014/12/4cc271adc7839d797e861f8940f5e06c.pdf

Tsunematsu, N., Yokoyama, H., Honjo, T., Ichihashi, A., Ando, H., Yamagata, Y., Murakami, D., & Shigyo, N. (2016). Relationship between land use variations and spatiotemporal changes in amounts of upward infrared radiative energy in downtown Tokyo at midday on hot summer days. *Annual report of the Tokyo Metropolitan Research Institute for Environmental Protection*, 76-82.

UN Habitat. (2016). *World cities report 2016: Urbanization and development: emerging futures*. Retrieved from http://wcr.unhabitat.org/main-report

UN. (2009). *Human Settlements Program. In Planning Sustainable Cities: Global Report on Human Settlements 2009*. London: Earthscan.

UNESCO. (2017). *Ocean literacy for all*. Retrieved from http://unesdoc.unesco.org/images/0026/002607/260721E.pdf

UNIDO. (2017). Accelerating clean energy through Industry 4.0: manufacturing the next revolution. A report of the United Nations Industrial Development Organization.

United Nations. (1992). Sustainable development. *Agenda 21*. Retrieved from https://sustainabledevelopment.un.org/outcomedocuments/agenda21

United Nations. (2016). *The World's Cities in 2016 Data Booklet*. Retrieved from http://www.un.org/en/development/desa/population/publications/pdf/urbanization/the_worlds_cities_in_2016_data_booklet.pdf

Uno, I., Wakamatsu, S., & Ueda, H. (1988). Behaviour of nocturnal urban boundary layer and air pollutants [in Japanese]. *Journal of Japan Society of Air Pollution*, *23*(2), 103–114.

Urbano, D. Y., & Díaz, J. C. (2009). Creación De Empresas E Instituciones: Un Modelo Teórico. In Creación De Empresas. Aproximación Al Estado Del Arte. Lisboa: Juruá.

Urteaga, E. (2013). *La teoría del capital social de Robert Putnam: Originalidad y carencias. Reflexión Política*. Retrieved from http://www.redalyc.org/articulo.oa?id=11028415005

Van Lancker, J., Wauters, E., & Van Huylenbroeck, G. (2016). Managing innovation in the bioeconomy: An open innovation perspective. *Biomass and Bioenergy*, *90*, 60–69. doi:10.1016/j.biombioe.2016.03.017

Vargas, J., Guerra, E., Bojórquez, A. Y., & Bojorquez, F. (2014). Gestión Estratégica De Organizaciones. Elaleph, 134.

Wakana, H. (2001). Development of contemporary fish apparel in Japan. *Water Resource and Environmental Research*, *14*, 1–9.

Waki, M., Sasaki, T., & Oshima, Y. (2015). A Dialogue analysis of "Exploration" in the Learning Cycle: Case of collaborative outdoor activity in "the Period of Integrated Study" at a junior high school. *Journal of Society of Clinical Education for Individual School Subjects*, *15*(2), 89–98.

Walker, B., Holling, C. S., Carpenter, S. R., & Kinzig, A. (2004). Resilience, Adaptability and Transformability in Social-Ecological Systems. *Ecology and Society*, *9*(2), 5. doi:10.5751/ES-00650-090205

Wang, Y., Schultz, S., & Giuf, I. F. (2008). Pictometry's proprietary airborne digital imaging system and its application in 3D city modelling. *International Archives of Photogrammetry, Sensing and Spatial Information Sciences,* (37), 1065-1069.

Weber, K. M., & Rohracher, H. (2012). Legitimizing research, technology and innovation policies for transformative change – Combining insights from innovation systems and multi-level perspective in a comprehensive 'failures' framework. *Research Policy*, *41*(6), 1037–1047. doi:10.1016/j.respol.2011.10.015

Wejchert, K. (1984). *Elementy kompozycji urbanistycznej*. Academic Press.

White House. (2012). *National Bioeconomy Blueprint*. Washington, DC: White House.

Wield, D., Hanlin, R., Mittra, J., & Smith, J. (2013). Twenty-first century bioeconomy: Global challenges of biological knowledge for health and agriculture. *Science & Public Policy*, *40*(1), 17–24. doi:10.1093cipolcs116

Wiesław. (n.d.). *Warszawskie pomniki*. Warszawa: Wydawnictwo "PTTK Kraj."

Winther, T. (2016). *Bioeconomy Strategies and Policies in the Baltic Sea Region Countries, State of Play*. Working Paper No. 1 of the Baltic Sea Regional Bioeconomy Council. Available online: http://bsrbioeconomy.net/resources/2016_docs/Working_Paper_1_%20BSR_Council.pdf

Wu, N., Wang, Q., & Xie, X. (2013). Wind energy harvesting with a piezoelectric harvester. *Smart Materials and Structures*, *22*(9), 095023. doi:10.1088/0964-1726/22/9/095023

Yamaguchi, Y. (2009). Forecast Population and Newspaper Report by Municipality in the Tohoku Region. *Quarterly Journal of Geography*, *61*(3), 234–238. doi:10.5190/tga.61.234

Yamashita & Matsumoto. (2014). Status of recycling plastic bottles in Japan and a comparison of the energy costs of different recycling methods. *International Journal of Environmental Protection and Policy*, *2*(4), 132-137.

Yamauchi, T., Okata, J., Koizumi, H., Murayama, A., & Manabe, R. (2015). A study of the process to determine public contribution elements in spatial districts for urban generation – Case studies in Ochanomizu, Shibuya, Ginza districts [in Japanese]. *Journal of the City Planning Institute of Japan*, *50*(3), 904–911.

Yung-Zielonko K. (2012). *Łączenie zaawansowanych i tradycyjnych technologii w architekturze proekologicznej"*. Oficyna Wydawnicza Politechniki Warszawskiej.

Zhang, T. (2014). Application of UAV Photography to Urban Planning Construction. *Urban Geotechnical Investigation & Surveying,* (2), 99-101.

Zilberman, D., Kim, E., Kirschner, S., Kaplan, S., & Reeves, J. (2014). Technology and the future bioeconomy. *Agricultural Economics*, *44*(s1), 95–102. doi:10.1111/agec.12054

Zinowiec-Cieplik, K. (2017). Potencjał integracji form roślinnych z architekturą - środowisko i technika. *Kwartalnik Naukowy Uczelni Vistula*, *4*(54), 2017.

Zuziak, Z. (2001). Strategies for Large Scale Urban Developments – the Game between the Public and Private Sector. In *Large Scale Urban Developments*. Academic Press.

Zuziak, Z.K. (2004). *Strategiczny wymiar urbanistyki.* Kraków: Politechniki Krakowskiej.

Related References

To continue our tradition of advancing academic research, we have compiled a list of recommended IGI Global readings. These references will provide additional information and guidance to further enrich your knowledge and assist you with your own research and future publications.

Abresch, J. (2003). Geographic Information Systems Research and Data Centers. In A. Hanson & B. Levin (Eds.), *Building a Virtual Library* (pp. 52–64). Hershey, PA: IGI Global. doi:10.4018/978-1-59140-106-3.ch004

Ajiboye, O. E., & Yusuff, O. S. (2017). Foreign Land Acquisition: Food Security and Food Chains – The Nigerian Experience. In I. Management Association (Ed.), Natural Resources Management: Concepts, Methodologies, Tools, and Applications (pp. 1524-1545). Hershey, PA: IGI Global. doi:10.4018/978-1-5225-0803-8.ch072

Alapiki, H. E., & Amadi, L. A. (2018). Sustainable Food Consumption in the Neoliberal Order: Challenges and Policy Implications. In A. Obayelu (Ed.), *Food Systems Sustainability and Environmental Policies in Modern Economies* (pp. 90–123). Hershey, PA: IGI Global. doi:10.4018/978-1-5225-3631-4.ch005

Altaş, A. (2018). Geographical Information System Applications Utilized in Museums in Turkey Within the Scope of the Cultural Heritage Tourism: A Case Study of Mobile Application of Müze Asist. In S. Chaudhuri & N. Ray (Eds.), *GIS Applications in the Tourism and Hospitality Industry* (pp. 42–60). Hershey, PA: IGI Global. doi:10.4018/978-1-5225-5088-4.ch002

Andreea, I. R. (2018). Beyond Macroeconomics of Food and Nutrition Security. *International Journal of Sustainable Economies Management, 7*(1), 13–22. doi:10.4018/IJSEM.2018010102

Anwar, J. (2017). Reproductive and Mental Health during Natural Disaster: Implications and Issues for Women in Developing Nations – A Case Example. In I. Management Association (Ed.), Gaming and Technology Addiction: Breakthroughs in Research and Practice (pp. 446-472). Hershey, PA: IGI Global. doi:10.4018/978-1-5225-0778-9.ch021

Awadh, H., Aksissou, M., Benhardouze, W., Darasi, F., & Snaiki, J. (2018). Socioeconomic Status of Artisanal Fishers in the West Part of Moroccan Mediterranean. *International Journal of Social Ecology and Sustainable Development, 9*(1), 40–52. doi:10.4018/IJSESD.2018010104

Aye, G. C., & Haruna, R. F. (2018). Effect of Climate Change on Crop Productivity and Prices in Benue State, Nigeria: Implications for Food Security. In V. Erokhin (Ed.), *Establishing Food Security and Alternatives to International Trade in Emerging Economies* (pp. 244–268). Hershey, PA: IGI Global. doi:10.4018/978-1-5225-2733-6.ch012

Azizan, S. A., & Suki, N. M. (2017). Consumers' Intentions to Purchase Organic Food Products. In T. Esakki (Ed.), *Green Marketing and Environmental Responsibility in Modern Corporations* (pp. 68–82). Hershey, PA: IGI Global. doi:10.4018/978-1-5225-2331-4.ch005

Barakabitze, A. A., Fue, K. G., Kitindi, E. J., & Sanga, C. A. (2017). Developing a Framework for Next Generation Integrated Agro Food-Advisory Systems in Developing Countries. In I. Management Association (Ed.), Agri-Food Supply Chain Management: Breakthroughs in Research and Practice (pp. 47-67). Hershey, PA: IGI Global. doi:10.4018/978-1-5225-1629-3.ch004

Beachcroft-Shaw, H., & Ellis, D. (2018). Using Successful Cases to Promote Environmental Sustainability: A Social Marketing Approach. In I. Management Association (Ed.), Sustainable Development: Concepts, Methodologies, Tools, and Applications (pp. 936-953). Hershey, PA: IGI Global. doi:10.4018/978-1-5225-3817-2.ch042

Behnassi, M., Kahime, K., Boussaa, S., Boumezzough, A., & Messouli, M. (2017). Infectious Diseases and Climate Vulnerability in Morocco: Governance and Adaptation Options. In I. Management Association (Ed.), Public Health and Welfare: Concepts, Methodologies, Tools, and Applications (pp. 91-109). Hershey, PA: IGI Global. doi:10.4018/978-1-5225-1674-3.ch005

Bekele, F., & Bekele, I. (2017). Social and Environmental Impacts on Agricultural Development. In W. Ganpat, R. Dyer, & W. Isaac (Eds.), *Agricultural Development and Food Security in Developing Nations* (pp. 21–56). Hershey, PA: IGI Global. doi:10.4018/978-1-5225-0942-4.ch002

Benaouda, A., & García-Peñalvo, F. J. (2018). Towards an Intelligent System for the Territorial Planning: Agricultural Case. In F. García-Peñalvo (Ed.), *Global Implications of Emerging Technology Trends* (pp. 158–178). Hershey, PA: IGI Global. doi:10.4018/978-1-5225-4944-4.ch010

Bhaskar, A., Rao, G. B., & Vencatesan, J. (2017). Characterization and Management Concerns of Water Resources around Pallikaranai Marsh, South Chennai. In P. Rao & Y. Patil (Eds.), *Reconsidering the Impact of Climate Change on Global Water Supply, Use, and Management* (pp. 102–121). Hershey, PA: IGI Global. doi:10.4018/978-1-5225-1046-8.ch007

Bianchini, E., & Sicurella, S. (2010). GIS: A New Tool for Criminology and Victimology's Studies. In R. Sette (Ed.), *Cases on Technologies for Teaching Criminology and Victimology: Methodologies and Practices* (pp. 87–110). Hershey, PA: IGI Global. doi:10.4018/978-1-60566-872-7.ch006

Bogataj, D., & Drobne, D. (2017). Control of Perishable Goods in Cold Logistic Chains by Bionanosensors. In S. Joo (Ed.), *Applying Nanotechnology for Environmental Sustainability* (pp. 376–402). Hershey, PA: IGI Global. doi:10.4018/978-1-5225-0585-3.ch016

Bogataj, D., & Drobne, D. (2017). Control of Perishable Goods in Cold Logistic Chains by Bionanosensors. In I. Management Association (Ed.), Materials Science and Engineering: Concepts, Methodologies, Tools, and Applications (pp. 471-497). Hershey, PA: IGI Global. doi:10.4018/978-1-5225-1798-6.ch019

Bogueva, D., & Marinova, D. (2018). What Is More Important: Perception of Masculinity or Personal Health and the Environment? In D. Bogueva, D. Marinova, & T. Raphaely (Eds.), *Handbook of Research on Social Marketing and Its Influence on Animal Origin Food Product Consumption* (pp. 148–162). Hershey, PA: IGI Global. doi:10.4018/978-1-5225-4757-0.ch010

Bouzid, M. (2017). Waterborne Diseases and Climate Change: Impact and Implications. In M. Bouzid (Ed.), *Examining the Role of Environmental Change on Emerging Infectious Diseases and Pandemics* (pp. 89–108). Hershey, PA: IGI Global. doi:10.4018/978-1-5225-0553-2.ch004

Bowles, D. C. (2017). Climate Change-Associated Conflict and Infectious Disease. In M. Bouzid (Ed.), *Examining the Role of Environmental Change on Emerging Infectious Diseases and Pandemics* (pp. 68–88). Hershey, PA: IGI Global. doi:10.4018/978-1-5225-0553-2.ch003

Buck, J. J., & Lowry, R. K. (2017). Oceanographic Data Management: Quills and Free Text to the Digital Age and "Big Data". In P. Diviacco, A. Leadbetter, & H. Glaves (Eds.), *Oceanographic and Marine Cross-Domain Data Management for Sustainable Development* (pp. 1–22). Hershey, PA: IGI Global. doi:10.4018/978-1-5225-0700-0.ch001

Buse, C. G. (2017). Are Climate Change Adaptation Policies a Game Changer?: A Case Study of Perspectives from Public Health Officials in Ontario, Canada. In M. Bouzid (Ed.), *Examining the Role of Environmental Change on Emerging Infectious Diseases and Pandemics* (pp. 230–257). Hershey, PA: IGI Global. doi:10.4018/978-1-5225-0553-2.ch010

Calderon, F. A., Giolo, E. G., Frau, C. D., Rengel, M. G., Rodriguez, H., Tornello, M., ... Gallucci, R. (2018). Seismic Microzonation and Site Effects Detection Through Microtremors Measures: A Review. In N. Ceryan (Ed.), *Handbook of Research on Trends and Digital Advances in Engineering Geology* (pp. 326–349). Hershey, PA: IGI Global. doi:10.4018/978-1-5225-2709-1.ch009

Carfì, D., Donato, A., & Panuccio, D. (2018). A Game Theory Coopetitive Perspective for Sustainability of Global Feeding: Agreements Among Vegan and Non-Vegan Food Firms. In I. Management Association (Ed.), Game Theory: Breakthroughs in Research and Practice (pp. 71-104). Hershey, PA: IGI Global. doi:10.4018/978-1-5225-2594-3.ch004

Carr, D. A., & Carr, T. R. (2007). Geographic Information System Applications in the Public Sector. In G. Garson (Ed.), *Modern Public Information Technology Systems: Issues and Challenges* (pp. 293–311). Hershey, PA: IGI Global. doi:10.4018/978-1-59904-051-6.ch015

Castagnolo, V. (2018). Analyzing, Classifying, Safeguarding: Drawing for the Borgo Murattiano Neighbourhood of Bari. In G. Carlone, N. Martinelli, & F. Rotondo (Eds.), *Designing Grid Cities for Optimized Urban Development and Planning* (pp. 93–108). Hershey, PA: IGI Global. doi:10.4018/978-1-5225-3613-0.ch006

Chekima, B. (2018). The Dilemma of Purchase Intention: A Conceptual Framework for Understanding Actual Consumption of Organic Food. *International Journal of Sustainable Economies Management*, *7*(2), 1–13. doi:10.4018/IJSEM.2018040101

Chen, Y. (2017). Sustainable Supply Chains and International Soft Landings: A Case of Wetland Entrepreneurship. In B. Christiansen & F. Kasarcı (Eds.), *Corporate Espionage, Geopolitics, and Diplomacy Issues in International Business* (pp. 232–247). Hershey, PA: IGI Global. doi:10.4018/978-1-5225-1031-4.ch013

Çıtak, L., Akel, V., & Ersoy, E. (2018). Investors' Reactions to the Announcement of New Constituents of BIST Sustainability Index: An Analysis by Event Study and Mean-Median Tests. In M. Risso & S. Testarmata (Eds.), *Value Sharing for Sustainable and Inclusive Development* (pp. 270–289). Hershey, PA: IGI Global. doi:10.4018/978-1-5225-3147-0.ch012

Crossland, M. D. (2009). Geographic Information Systems as Decision Tools. In M. Khosrow-Pour, D.B.A. (Ed.), Encyclopedia of Information Science and Technology, Second Edition (pp. 1630-1633). Hershey, PA: IGI Global. doi:10.4018/978-1-60566-026-4.ch257

Crossland, M. D., Herschel, R. T., Perkins, W. C., & Scudder, J. N. (2002). Geographic Information Systems: How Cognitive Style Impacts Decision-Making Effectiveness. In M. Mahmood (Ed.), *Advanced Topics in End User Computing* (Vol. 1, pp. 285–301). Hershey, PA: IGI Global. doi:10.4018/978-1-930708-42-6.ch017

D'Aleo, V., D'Aleo, F., & Bonanno, R. (2018). New Food Industries Toward a New Level of Sustainable Supply: Success Stories, Business Models, and Strategies. In V. Erokhin (Ed.), *Establishing Food Security and Alternatives to International Trade in Emerging Economies* (pp. 74–97). Hershey, PA: IGI Global. doi:10.4018/978-1-5225-2733-6.ch004

Dagevos, H., & Reinders, M. J. (2018). Flexitarianism and Social Marketing: Reflections on Eating Meat in Moderation. In D. Bogueva, D. Marinova, & T. Raphaely (Eds.), *Handbook of Research on Social Marketing and Its Influence on Animal Origin Food Product Consumption* (pp. 105–120). Hershey, PA: IGI Global. doi:10.4018/978-1-5225-4757-0.ch007

Deenapanray, P. N., & Ramma, I. (2017). Adaptations to Climate Change and Climate Variability in the Agriculture Sector in Mauritius: Lessons from a Technical Needs Assessment. In I. Management Association (Ed.), Natural Resources Management: Concepts, Methodologies, Tools, and Applications (pp. 655-680). Hershey, PA: IGI Global. doi:10.4018/978-1-5225-0803-8.ch030

Deenapanray, P. N., & Ramma, I. (2017). Adaptations to Climate Change and Climate Variability in the Agriculture Sector in Mauritius: Lessons from a Technical Needs Assessment. In I. Management Association (Ed.), Natural Resources Management: Concepts, Methodologies, Tools, and Applications (pp. 655-680). Hershey, PA: IGI Global. doi:10.4018/978-1-5225-0803-8.ch030

Deshpande, S., Basu, S. K., Li, X., & Chen, X. (2017). Smart, Innovative and Intelligent Technologies Used in Drug Designing. In I. Management Association (Ed.), Pharmaceutical Sciences: Breakthroughs in Research and Practice (pp. 1175-1191). Hershey, PA: IGI Global. doi:10.4018/978-1-5225-1762-7.ch045

Dlamini, P. N. (2017). Use of Information and Communication Technologies Tools to Capture, Store, and Disseminate Indigenous Knowledge: A Literature Review. In P. Ngulube (Ed.), *Handbook of Research on Theoretical Perspectives on Indigenous Knowledge Systems in Developing Countries* (pp. 225–247). Hershey, PA: IGI Global. doi:10.4018/978-1-5225-0833-5.ch010

Dolejsova, M., & Kera, D. (2017). The Fermentation GutHub Project and the Internet of Microbes. In S. Konomi & G. Roussos (Eds.), *Enriching Urban Spaces with Ambient Computing, the Internet of Things, and Smart City Design* (pp. 25–46). Hershey, PA: IGI Global. doi:10.4018/978-1-5225-0827-4.ch002

Dolunay, O. (2018). A Paradigm Shift: Empowering Farmers to Eliminate the Waste in the Form of Fresh Water and Energy Through the Implementation of 4R+T. In I. Management Association (Ed.), Sustainable Development: Concepts, Methodologies, Tools, and Applications (pp. 882-892). Hershey, PA: IGI Global. doi:10.4018/978-1-5225-3817-2.ch039

Dube, P., Heijman, W. J., Ihle, R., & Ochieng, J. (2018). The Potential of Traditional Leafy Vegetables for Improving Food Security in Africa. In V. Erokhin (Ed.), *Establishing Food Security and Alternatives to International Trade in Emerging Economies* (pp. 220–243). Hershey, PA: IGI Global. doi:10.4018/978-1-5225-2733-6.ch011

Duruji, M. M., & Urenma, D. F. (2017). The Environmentalism and Politics of Climate Change: A Study of the Process of Global Convergence through UNFCCC Conferences. In I. Management Association (Ed.), *Natural Resources Management: Concepts, Methodologies, Tools, and Applications* (pp. 77-108). Hershey, PA: IGI Global. 10.4018/978-1-5225-0803-8.ch004

Dutta, U. (2017). Agro-Geoinformatics, Potato Cultivation, and Climate Change. In S. Londhe (Ed.), *Sustainable Potato Production and the Impact of Climate Change* (pp. 247–271). Hershey, PA: IGI Global. doi:10.4018/978-1-5225-1715-3.ch012

Edirisinghe, R., Stranieri, A., & Wickramasinghe, N. (2017). A Taxonomy for mHealth. In N. Wickramasinghe (Ed.), *Handbook of Research on Healthcare Administration and Management* (pp. 596–615). Hershey, PA: IGI Global. doi:10.4018/978-1-5225-0920-2.ch036

Ekpeni, N. M., & Ayeni, A. O. (2018). Global Natural Hazard and Disaster Vulnerability Management. In A. Eneanya (Ed.), *Handbook of Research on Environmental Policies for Emergency Management and Public Safety* (pp. 83–104). Hershey, PA: IGI Global. doi:10.4018/978-1-5225-3194-4.ch005

Ene, C., Voica, M. C., & Panait, M. (2017). Green Investments and Food Security: Opportunities and Future Directions in the Context of Sustainable Development. In M. Mieila (Ed.), *Measuring Sustainable Development and Green Investments in Contemporary Economies* (pp. 163–200). Hershey, PA: IGI Global. doi:10.4018/978-1-5225-2081-8.ch007

Escamilla, I., Ruíz, M. T., Ibarra, M. M., Soto, V. L., Quintero, R., & Guzmán, G. (2018). Geocoding Tweets Based on Semantic Web and Ontologies. In M. Lytras, N. Aljohani, E. Damiani, & K. Chui (Eds.), *Innovations, Developments, and Applications of Semantic Web and Information Systems* (pp. 372–392). Hershey, PA: IGI Global. doi:10.4018/978-1-5225-5042-6.ch014

Escribano, A. J. (2018). Marketing Strategies for Trendy Animal Products: Sustainability as a Core. In F. Quoquab, R. Thurasamy, & J. Mohammad (Eds.), *Driving Green Consumerism Through Strategic Sustainability Marketing* (pp. 169–203). Hershey, PA: IGI Global. doi:10.4018/978-1-5225-2912-5.ch010

Eudoxie, G., & Roopnarine, R. (2017). Climate Change Adaptation and Disaster Risk Management in the Caribbean. In W. Ganpat & W. Isaac (Eds.), *Environmental Sustainability and Climate Change Adaptation Strategies* (pp. 97–125). Hershey, PA: IGI Global. doi:10.4018/978-1-5225-1607-1.ch004

Farmer, L. S. (2017). Data Analytics for Strategic Management: Getting the Right Data. In V. Wang (Ed.), *Encyclopedia of Strategic Leadership and Management* (pp. 810–822). Hershey, PA: IGI Global. doi:10.4018/978-1-5225-1049-9.ch056

Fattal, L. R. (2017). Catastrophe: An Uncanny Catalyst for Creativity. In R. Shin (Ed.), *Convergence of Contemporary Art, Visual Culture, and Global Civic Engagement* (pp. 244–262). Hershey, PA: IGI Global. doi:10.4018/978-1-5225-1665-1.ch014

Forti, I. (2017). A Cross Reading of Landscape through Digital Landscape Models: The Case of Southern Garda. In A. Ippolito (Ed.), *Handbook of Research on Emerging Technologies for Architectural and Archaeological Heritage* (pp. 532–561). Hershey, PA: IGI Global. doi:10.4018/978-1-5225-0675-1.ch018

Gharbi, A., De Runz, C., & Akdag, H. (2017). Urban Development Modelling: A Survey. In S. Faiz & K. Mahmoudi (Eds.), *Handbook of Research on Geographic Information Systems Applications and Advancements* (pp. 96–124). Hershey, PA: IGI Global. doi:10.4018/978-1-5225-0937-0.ch004

Ghosh, I., & Ghoshal, I. (2018). Implications of Trade Liberalization for Food Security Under the ASEAN-India Strategic Partnership: A Gravity Model Approach. In V. Erokhin (Ed.), *Establishing Food Security and Alternatives to International Trade in Emerging Economies* (pp. 98–118). Hershey, PA: IGI Global. doi:10.4018/978-1-5225-2733-6.ch005

Glaves, H. M. (2017). Developing a Common Global Framework for Marine Data Management. In P. Diviacco, A. Leadbetter, & H. Glaves (Eds.), *Oceanographic and Marine Cross-Domain Data Management for Sustainable Development* (pp. 47–68). Hershey, PA: IGI Global. doi:10.4018/978-1-5225-0700-0.ch003

Godulla, A., & Wolf, C. (2018). Future of Food: Transmedia Strategies of National Geographic. In R. Gambarato & G. Alzamora (Eds.), *Exploring Transmedia Journalism in the Digital Age* (pp. 162–182). Hershey, PA: IGI Global. doi:10.4018/978-1-5225-3781-6.ch010

Gomes, P. P. (2018). Food and Environment: A Review on the Sustainability of Six Different Dietary Patterns. In A. Obayelu (Ed.), *Food Systems Sustainability and Environmental Policies in Modern Economies* (pp. 15–31). Hershey, PA: IGI Global. doi:10.4018/978-1-5225-3631-4.ch002

Gonzalez-Feliu, J. (2018). Sustainability Evaluation of Green Urban Logistics Systems: Literature Overview and Proposed Framework. In A. Paul, D. Bhattacharyya, & S. Anand (Eds.), *Green Initiatives for Business Sustainability and Value Creation* (pp. 103–134). Hershey, PA: IGI Global. doi:10.4018/978-1-5225-2662-9.ch005

Goodland, R. (2017). A Fresh Look at Livestock Greenhouse Gas Emissions and Mitigation. In I. Management Association (Ed.), Natural Resources Management: Concepts, Methodologies, Tools, and Applications (pp. 124-139). Hershey, PA: IGI Global. doi:10.4018/978-1-5225-0803-8.ch006

Goundar, S., & Appana, S. (2018). Mainstreaming Development Policies for Climate Change in Fiji: A Policy Gap Analysis and the Role of ICTs. In I. Management Association (Ed.), Sustainable Development: Concepts, Methodologies, Tools, and Applications (pp. 402-432). Hershey, PA: IGI Global. doi:10.4018/978-1-5225-3817-2.ch020

Granell-Canut, C., & Aguilar-Moreno, E. (2018). Geospatial Influence in Science Mapping. In M. Khosrow-Pour, D.B.A. (Ed.), Encyclopedia of Information Science and Technology, Fourth Edition (pp. 3473-3483). Hershey, PA: IGI Global. doi:10.4018/978-1-5225-2255-3.ch302

Grigelis, A., Blažauskas, N., Gelumbauskaitė, L. Ž., Gulbinskas, S., Suzdalev, S., & Ferrarin, C. (2017). Marine Environment Data Management Related to the Human Activity in the South-Eastern Baltic Sea (The Lithuanian Sector). In P. Diviacco, A. Leadbetter, & H. Glaves (Eds.), *Oceanographic and Marine Cross-Domain Data Management for Sustainable Development* (pp. 282–302). Hershey, PA: IGI Global. doi:10.4018/978-1-5225-0700-0.ch012

Guma, I. P., Rwashana, A. S., & Oyo, B. (2018). Food Security Policy Analysis Using System Dynamics: The Case of Uganda. *International Journal of Information Technologies and Systems Approach, 11*(1), 72–90. doi:10.4018/IJITSA.2018010104

Guma, I. P., Rwashana, A. S., & Oyo, B. (2018). Food Security Indicators for Subsistence Farmers Sustainability: A System Dynamics Approach. *International Journal of System Dynamics Applications*, *7*(1), 45–64. doi:10.4018/IJSDA.2018010103

Gupta, P., & Goyal, S. (2017). Wildlife Habitat Evaluation. In A. Santra & S. Mitra (Eds.), *Remote Sensing Techniques and GIS Applications in Earth and Environmental Studies* (pp. 258–264). Hershey, PA: IGI Global. doi:10.4018/978-1-5225-1814-3.ch013

Hanson, T., & Hildebrand, E. (2018). GPS Travel Diaries in Rural Transportation Research: A Focus on Older Drivers. In I. Management Association (Ed.), Intelligent Transportation and Planning: Breakthroughs in Research and Practice (pp. 609-625). Hershey, PA: IGI Global. doi:10.4018/978-1-5225-5210-9.ch027

Hartman, M. B. (2017). Research-Based Climate Change Public Education Programs. In I. Management Association (Ed.), Natural Resources Management: Concepts, Methodologies, Tools, and Applications (pp. 992-1003). Hershey, PA: IGI Global. doi:10.4018/978-1-5225-0803-8.ch046

Hashim, N. (2017). Zanzibari Seaweed: Global Climate Change and the Promise of Adaptation. In I. Management Association (Ed.), Natural Resources Management: Concepts, Methodologies, Tools, and Applications (pp. 365-391). Hershey, PA: IGI Global. doi:10.4018/978-1-5225-0803-8.ch019

Herrera, J. E., Argüello, L. V., Gonzalez-Feliu, J., & Jaimes, W. A. (2017). Decision Support System Design Requirements, Information Management, and Urban Logistics Efficiency: Case Study of Bogotá, Colombia. In G. Jamil, A. Soares, & C. Pessoa (Eds.), *Handbook of Research on Information Management for Effective Logistics and Supply Chains* (pp. 223–238). Hershey, PA: IGI Global. doi:10.4018/978-1-5225-0973-8.ch012

Huizinga, T., Ayanso, A., Smoor, M., & Wronski, T. (2017). Exploring Insurance and Natural Disaster Tweets Using Text Analytics. *International Journal of Business Analytics*, *4*(1), 1–17. doi:10.4018/IJBAN.2017010101

Hung, K., Kalantari, M., & Rajabifard, A. (2017). An Integrated Method for Assessing the Text Content Quality of Volunteered Geographic Information in Disaster Management. *International Journal of Information Systems for Crisis Response and Management*, *9*(2), 1–17. doi:10.4018/IJISCRAM.2017040101

Husnain, A., & Avdic, A. (2018). Identifying the Contemporary Status of E-Service Sustainability Research. In I. Management Association (Ed.), Sustainable Development: Concepts, Methodologies, Tools, and Applications (pp. 467-485). Hershey, PA: IGI Global. doi:10.4018/978-1-5225-3817-2.ch022

Iarossi, M. P., & Ferro, L. (2017). "The Past is Never Dead. It's Not Even Past": Virtual Archaeological Promenade. In A. Ippolito & M. Cigola (Eds.), *Handbook of Research on Emerging Technologies for Digital Preservation and Information Modeling* (pp. 228–255). Hershey, PA: IGI Global. doi:10.4018/978-1-5225-0680-5.ch010

Ignjatijević, S., & Cvijanović, D. (2018). Analysis of Serbian Honey Production and Exports. In *Exploring the Global Competitiveness of Agri-Food Sectors and Serbia's Dominant Presence: Emerging Research and Opportunities* (pp. 109–139). Hershey, PA: IGI Global. doi:10.4018/978-1-5225-2762-6.ch005

Jana, S. K., & Karmakar, A. K. (2017). Globalization, Governance, and Food Security: The Case of BRICS. In I. Management Association (Ed.), Natural Resources Management: Concepts, Methodologies, Tools, and Applications (pp. 692-712). Hershey, PA: IGI Global. doi:10.4018/978-1-5225-0803-8.ch032

Jana, S. K., & Karmakar, A. K. (2017). Food Security in Asia: Is There Convergence? In I. Management Association (Ed.), Natural Resources Management: Concepts, Methodologies, Tools, and Applications (pp. 109-123). Hershey, PA: IGI Global. doi:10.4018/978-1-5225-0803-8.ch005

John, J., & Kumar, S. (2018). A Locational Decision Making Framework for Shipbreaking Under Multiple Criteria. In I. Management Association (Ed.), Operations and Service Management: Concepts, Methodologies, Tools, and Applications (pp. 504-527). Hershey, PA: IGI Global. doi:10.4018/978-1-5225-3909-4.ch024

John, J., & Srivastava, R. K. (2018). Decision Insights for Shipbreaking using Environmental Impact Assessment: Review and Perspectives. *International Journal of Strategic Decision Sciences*, *9*(1), 45–62. doi:10.4018/IJSDS.2018010104

Joshi, Y., & Rahman, Z. (2018). Determinants of Sustainable Consumption Behaviour: Review and Conceptual Framework. In A. Paul, D. Bhattacharyya, & S. Anand (Eds.), *Green Initiatives for Business Sustainability and Value Creation* (pp. 239–262). Hershey, PA: IGI Global. doi:10.4018/978-1-5225-2662-9.ch011

Juma, D. W., Reuben, M., Wang, H., & Li, F. (2018). Adaptive Coevolution: Realigning the Water Governance Regime to the Changing Climate. In I. Management Association (Ed.), Hydrology and Water Resource Management: Breakthroughs in Research and Practice (pp. 346-357). Hershey, PA: IGI Global. doi:10.4018/978-1-5225-3427-3.ch014

K., S., & Tripathy, B. K. (2018). Neighborhood Rough-Sets-Based Spatial Data Analytics. In M. Khosrow-Pour, D.B.A. (Ed.), *Encyclopedia of Information Science and Technology, Fourth Edition* (pp. 1835-1844). Hershey, PA: IGI Global. doi:10.4018/978-1-5225-2255-3.ch160

Kabir, F. (2018). Towards a More Gender-Inclusive Climate Change Policy. In N. Mahtab, T. Haque, I. Khan, M. Islam, & I. Wahid (Eds.), *Handbook of Research on Women's Issues and Rights in the Developing World* (pp. 354–369). Hershey, PA: IGI Global. doi:10.4018/978-1-5225-3018-3.ch022

Kabir, F. (2018). Towards a More Gender-Inclusive Climate Change Policy. In I. Management Association (Ed.), Climate Change and Environmental Concerns: Breakthroughs in Research and Practice (pp. 525-540). Hershey, PA: IGI Global. doi:10.4018/978-1-5225-5487-5.ch027

Kanyamuka, J. S., Jumbe, C. B., & Ricker-Gilbert, J. (2018). Making Agricultural Input Subsidies More Effective and Profitable in Africa: The Role of Complementary Interventions. In A. Obayelu (Ed.), *Food Systems Sustainability and Environmental Policies in Modern Economies* (pp. 172–187). Hershey, PA: IGI Global. doi:10.4018/978-1-5225-3631-4.ch008

Karimi, H., & Gholamrezafahimi, F. (2017). Study of Integrated Coastal Zone Management and Its Environmental Effects: A Case of Iran. In R. Singh, A. Singh, & V. Srivastava (Eds.), *Environmental Issues Surrounding Human Overpopulation* (pp. 64–88). Hershey, PA: IGI Global. doi:10.4018/978-1-5225-1683-5.ch004

Kaya, I. R., Hutabarat, J., & Bambang, A. N. (2018). "Sasi": A New Path to Sustain Seaweed Farming From Up-Stream to Down-Stream in Kotania Bay, Molucass. *International Journal of Social Ecology and Sustainable Development*, *9*(2), 28–36. doi:10.4018/IJSESD.2018040103

Khader, V. (2018). Technologies for Food, Health, Livelihood, and Nutrition Security. In I. Management Association (Ed.), Food Science and Nutrition: Breakthroughs in Research and Practice (pp. 94-112). Hershey, PA: IGI Global. doi:10.4018/978-1-5225-5207-9.ch005

Kocadağlı, A. Y. (2017). The Temporal and Spatial Development of Organic Agriculture in Turkey. In W. Ganpat, R. Dyer, & W. Isaac (Eds.), *Agricultural Development and Food Security in Developing Nations* (pp. 130–156). Hershey, PA: IGI Global. doi:10.4018/978-1-5225-0942-4.ch006

Koundouri, P., Giannouli, A., & Souliotis, I. (2017). An Integrated Approach for Sustainable Environmental and Socio-Economic Development Using Offshore Infrastructure. In I. Management Association (Ed.), Renewable and Alternative Energy: Concepts, Methodologies, Tools, and Applications (pp. 1581-1601). Hershey, PA: IGI Global. doi:10.4018/978-1-5225-1671-2.ch056

Kumar, A., & Dash, M. K. (2017). Sustainability and Future Generation Infrastructure on Digital Platform: A Study of Generation Y. In N. Ray (Ed.), *Business Infrastructure for Sustainability in Developing Economies* (pp. 124–142). Hershey, PA: IGI Global. doi:10.4018/978-1-5225-2041-2.ch007

Kumar, A., Mukherjee, A. B., & Krishna, A. P. (2017). Application of Conventional Data Mining Techniques and Web Mining to Aid Disaster Management. In A. Kumar (Ed.), *Web Usage Mining Techniques and Applications Across Industries* (pp. 138–167). Hershey, PA: IGI Global. doi:10.4018/978-1-5225-0613-3.ch006

Kumar, C. P. (2017). Impact of Climate Change on Groundwater Resources. In I. Management Association (Ed.), Natural Resources Management: Concepts, Methodologies, Tools, and Applications (pp. 1094-1120). Hershey, PA: IGI Global. doi:10.4018/978-1-5225-0803-8.ch052

Kumari, S., & Patil, Y. (2017). Achieving Climate Smart Agriculture with a Sustainable Use of Water: A Conceptual Framework for Sustaining the Use of Water for Agriculture in the Era of Climate Change. In P. Rao & Y. Patil (Eds.), *Reconsidering the Impact of Climate Change on Global Water Supply, Use, and Management* (pp. 122–143). Hershey, PA: IGI Global. doi:10.4018/978-1-5225-1046-8.ch008

Kumari, S., & Patil, Y. (2018). Achieving Climate Smart Agriculture With a Sustainable Use of Water: A Conceptual Framework for Sustaining the Use of Water for Agriculture in the Era of Climate Change. In I. Management Association (Ed.), Climate Change and Environmental Concerns: Breakthroughs in Research and Practice (pp. 111-133). Hershey, PA: IGI Global. doi:10.4018/978-1-5225-5487-5.ch006

Kursah, M. B. (2017). Least-Cost Pipeline using Geographic Information System: The Limit to Technicalities. *International Journal of Applied Geospatial Research, 8*(3), 1–15. doi:10.4018/ijagr.2017070101

Lahiri, S., Ghosh, D., & Bhakta, J. N. (2017). Role of Microbes in Eco-Remediation of Perturbed Aquatic Ecosystem. In J. Bhakta (Ed.), *Handbook of Research on Inventive Bioremediation Techniques* (pp. 70–107). Hershey, PA: IGI Global. doi:10.4018/978-1-5225-2325-3.ch004

Lallo, C. H., Smalling, S., Facey, A., & Hughes, M. (2018). The Impact of Climate Change on Small Ruminant Performance in Caribbean Communities. In I. Management Association (Ed.), Climate Change and Environmental Concerns: Breakthroughs in Research and Practice (pp. 193-218). Hershey, PA: IGI Global. doi:10.4018/978-1-5225-5487-5.ch010

Laurini, R. (2017). Nature of Geographic Knowledge Bases. In S. Faiz & K. Mahmoudi (Eds.), *Handbook of Research on Geographic Information Systems Applications and Advancements* (pp. 29–60). Hershey, PA: IGI Global. doi:10.4018/978-1-5225-0937-0.ch002

Lawrence, J., Simpson, L., & Piggott, A. (2017). Protected Agriculture: A Climate Change Adaptation for Food and Nutrition Security. In I. Management Association (Ed.), Natural Resources Management: Concepts, Methodologies, Tools, and Applications (pp. 140-158). Hershey, PA: IGI Global. doi:10.4018/978-1-5225-0803-8.ch007

Leadbetter, A., Cheatham, M., Shepherd, A., & Thomas, R. (2017). Linked Ocean Data 2.0. In P. Diviacco, A. Leadbetter, & H. Glaves (Eds.), *Oceanographic and Marine Cross-Domain Data Management for Sustainable Development* (pp. 69–99). Hershey, PA: IGI Global. doi:10.4018/978-1-5225-0700-0.ch004

Leipnik, M. R., & Mehta, S. S. (2005). Geographic Information Systems (GIS) in E-Marketing. In I. Clarke III & T. Flaherty (Eds.), *Advances in Electronic Marketing* (pp. 193–210). Hershey, PA: IGI Global. doi:10.4018/978-1-59140-321-0.ch011

Lucas, M. R., Rego, C., Vieira, C., & Vieira, I. (2017). Proximity and Cooperation for Innovative Regional Development: The Case of the Science and Technology Park of Alentejo. In L. Carvalho (Ed.), *Handbook of Research on Entrepreneurial Development and Innovation Within Smart Cities* (pp. 199–228). Hershey, PA: IGI Global. doi:10.4018/978-1-5225-1978-2.ch010

Ma, X., Beaulieu, S. E., Fu, L., Fox, P., Di Stefano, M., & West, P. (2017). Documenting Provenance for Reproducible Marine Ecosystem Assessment in Open Science. In P. Diviacco, A. Leadbetter, & H. Glaves (Eds.), *Oceanographic and Marine Cross-Domain Data Management for Sustainable Development* (pp. 100–126). Hershey, PA: IGI Global. doi:10.4018/978-1-5225-0700-0.ch005

Mabe, L. K., & Oladele, O. I. (2017). Application of Information Communication Technologies for Agricultural Development through Extension Services: A Review. In T. Tossy (Ed.), *Information Technology Integration for Socio-Economic Development* (pp. 52–101). Hershey, PA: IGI Global. doi:10.4018/978-1-5225-0539-6.ch003

Malomo, B. I. (2018). A Review of Psychological Resilience as a Response to Natural Hazards in Nigeria. In A. Eneanya (Ed.), *Handbook of Research on Environmental Policies for Emergency Management and Public Safety* (pp. 147–165). Hershey, PA: IGI Global. doi:10.4018/978-1-5225-3194-4.ch008

Manchiraju, S. (2018). Predicting Behavioral Intentions Toward Sustainable Fashion Consumption: A Comparison of Attitude-Behavior and Value-Behavior Consistency Models. In I. Management Association (Ed.), Fashion and Textiles: Breakthroughs in Research and Practice (pp. 1-21). Hershey, PA: IGI Global. doi:10.4018/978-1-5225-3432-7.ch001

Manzella, G. M., Bartolini, R., Bustaffa, F., D'Angelo, P., De Mattei, M., Frontini, F., ... Spada, A. (2017). Semantic Search Engine for Data Management and Sustainable Development: Marine Planning Service Platform. In P. Diviacco, A. Leadbetter, & H. Glaves (Eds.), *Oceanographic and Marine Cross-Domain Data Management for Sustainable Development* (pp. 127–154). Hershey, PA: IGI Global. doi:10.4018/978-1-5225-0700-0.ch006

Mbonigaba, J. (2018). Comparing the Effects of Unsustainable Production and Consumption of Food on Health and Policy Across Developed and Less Developed Countries. In A. Obayelu (Ed.), *Food Systems Sustainability and Environmental Policies in Modern Economies* (pp. 124–158). Hershey, PA: IGI Global. doi:10.4018/978-1-5225-3631-4.ch006

McKeown, A. E. (2017). Nurses, Healthcare, and Environmental Pollution and Solutions: Breaking the Cycle of Harm. In I. Management Association (Ed.), Natural Resources Management: Concepts, Methodologies, Tools, and Applications (pp. 392-415). Hershey, PA: IGI Global. doi:10.4018/978-1-5225-0803-8.ch020

Mennecke, B. E., & West, L. A. Jr. (2001). Geographic Information Systems in Developing Countries: Issues in Data Collection, Implementation and Management. *Journal of Global Information Management*, *9*(4), 44–54. doi:10.4018/jgim.2001100103

Mili, B., Barua, A., & Katyaini, S. (2017). Climate Change and Adaptation through the Lens of Capability Approach: A Case Study from Darjeeling, Eastern Himalaya. In I. Management Association (Ed.), Natural Resources Management: Concepts, Methodologies, Tools, and Applications (pp. 1351-1365). Hershey, PA: IGI Global. doi:10.4018/978-1-5225-0803-8.ch064

Mir, S. A., Shah, M. A., Mir, M. M., & Iqbal, U. (2017). New Horizons of Nanotechnology in Agriculture and Food Processing Industry. In B. Nayak, A. Nanda, & M. Bhat (Eds.), *Integrating Biologically-Inspired Nanotechnology into Medical Practice* (pp. 230–258). Hershey, PA: IGI Global. doi:10.4018/978-1-5225-0610-2.ch009

Mujere, N., & Moyce, W. (2017). Climate Change Impacts on Surface Water Quality. In W. Ganpat & W. Isaac (Eds.), *Environmental Sustainability and Climate Change Adaptation Strategies* (pp. 322–340). Hershey, PA: IGI Global. doi:10.4018/978-1-5225-1607-1.ch012

Mukherjee, A. B., Krishna, A. P., & Patel, N. (2018). Geospatial Technology for Urban Sciences. In *Geospatial Technologies in Urban System Development: Emerging Research and Opportunities* (pp. 99–120). Hershey, PA: IGI Global. doi:10.4018/978-1-5225-3683-3.ch005

Nagarajan, S. K., & Sangaiah, A. K. (2017). Vegetation Index: Ideas, Methods, Influences, and Trends. In N. Kumar, A. Sangaiah, M. Arun, & S. Anand (Eds.), *Advanced Image Processing Techniques and Applications* (pp. 347–386). Hershey, PA: IGI Global. doi:10.4018/978-1-5225-2053-5.ch016

Naraine, L., & Meehan, K. (2017). Strengthening Food Security with Sustainable Practices by Smallholder Farmers in Lesser Developed Economies. In W. Ganpat, R. Dyer, & W. Isaac (Eds.), *Agricultural Development and Food Security in Developing Nations* (pp. 57–81). Hershey, PA: IGI Global. doi:10.4018/978-1-5225-0942-4.ch003

Naraine, L., & Meehan, K. (2017). Strengthening Food Security with Sustainable Practices by Smallholder Farmers in Lesser Developed Economies. In W. Ganpat, R. Dyer, & W. Isaac (Eds.), *Agricultural Development and Food Security in Developing Nations* (pp. 57–81). Hershey, PA: IGI Global. doi:10.4018/978-1-5225-0942-4.ch003

Naraine, L., & Meehan, K. (2017). Strengthening Food Security with Sustainable Practices by Smallholder Farmers in Lesser Developed Economies. In W. Ganpat, R. Dyer, & W. Isaac (Eds.), *Agricultural Development and Food Security in Developing Nations* (pp. 57–81). Hershey, PA: IGI Global. doi:10.4018/978-1-5225-0942-4.ch003

Neufeld, D. J., & Griffith, S. (2006). Isobord's Geographic Information System (GIS) Solution. In M. Khosrow-Pour, D.B.A. (Ed.), Cases on Database Technologies and Applications (pp. 64-80). Hershey, PA: IGI Global. doi:10.4018/978-1-59904-399-9.ch004

Nikolaou, K., Tsakiridou, E., Anastasiadis, F., & Mattas, K. (2017). Exploring Alternative Distribution Channels of Agricultural Products. *International Journal of Food and Beverage Manufacturing and Business Models*, *2*(2), 36–66. doi:10.4018/IJFBMBM.2017070103

Nishat, K. J., & Rahman, M. S. (2018). Disaster, Vulnerability, and Violence Against Women: Global Findings and a Research Agenda for Bangladesh. In N. Mahtab, T. Haque, I. Khan, M. Islam, & I. Wahid (Eds.), *Handbook of Research on Women's Issues and Rights in the Developing World* (pp. 235–250). Hershey, PA: IGI Global. doi:10.4018/978-1-5225-3018-3.ch014

O'Hara, S., Jones, D., & Trobman, H. B. (2018). Building an Urban Food System Through UDC Food Hubs. In A. Burtin, J. Fleming, & P. Hampton-Garland (Eds.), *Changing Urban Landscapes Through Public Higher Education* (pp. 116–143). Hershey, PA: IGI Global. doi:10.4018/978-1-5225-3454-9.ch006

Obayelu, A. E. (2018). Integrating Environment, Food Systems, and Sustainability in Feeding the Growing Population in Developing Countries. In A. Obayelu (Ed.), *Food Systems Sustainability and Environmental Policies in Modern Economies* (pp. 1–14). Hershey, PA: IGI Global. doi:10.4018/978-1-5225-3631-4.ch001

Othman, R., Nath, N., & Laswad, F. (2018). Environmental Reporting and Accounting: Sustainability Hybridisation. In G. Azevedo, J. da Silva Oliveira, R. Marques, & A. Ferreira (Eds.), *Handbook of Research on Modernization and Accountability in Public Sector Management* (pp. 130–158). Hershey, PA: IGI Global. doi:10.4018/978-1-5225-3731-1.ch007

Padigala, B. S. (2018). Traditional Water Management System for Climate Change Adaptation in Mountain Ecosystems. In I. Management Association (Ed.), Climate Change and Environmental Concerns: Breakthroughs in Research and Practice (pp. 630-655). Hershey, PA: IGI Global. doi:10.4018/978-1-5225-5487-5.ch033

Panda, C. K. (2018). Mobile Phone Usage in Agricultural Extension in India: The Current and Future Perspective. In F. Mtenzi, G. Oreku, D. Lupiana, & J. Yonazi (Eds.), *Mobile Technologies and Socio-Economic Development in Emerging Nations* (pp. 1–21). Hershey, PA: IGI Global. doi:10.4018/978-1-5225-4029-8.ch001

Pandian, S. L., Yarrakula, K., & Chaudhury, P. (2018). GIS-Based Decision Support System for Village Level: A Case Study in Andhra Pradesh. In S. Chaudhuri & N. Ray (Eds.), *GIS Applications in the Tourism and Hospitality Industry* (pp. 275–295). Hershey, PA: IGI Global. doi:10.4018/978-1-5225-5088-4.ch012

Quaranta, G., & Salvia, R. (2017). Social-Based Product Innovation and Governance in The Milk Sector: The Case of Carciocacio and Innonatura. In T. Tarnanidis, M. Vlachopoulou, & J. Papathanasiou (Eds.), *Driving Agribusiness With Technology Innovations* (pp. 293–310). Hershey, PA: IGI Global. doi:10.4018/978-1-5225-2107-5.ch015

Rahman, M. K., Schmidlin, T. W., Munro-Stasiuk, M. J., & Curtis, A. (2017). Geospatial Analysis of Land Loss, Land Cover Change, and Landuse Patterns of Kutubdia Island, Bangladesh. *International Journal of Applied Geospatial Research*, *8*(2), 45–60. doi:10.4018/IJAGR.2017040104

Rajack-Talley, T. A. (2017). Agriculture, Trade Liberalization and Poverty in the ACP Countries. In W. Ganpat, R. Dyer, & W. Isaac (Eds.), *Agricultural Development and Food Security in Developing Nations* (pp. 1–20). Hershey, PA: IGI Global. doi:10.4018/978-1-5225-0942-4.ch001

Rajamanickam, S. (2018). Exploring Landscapes in Regional Convergence: Environment and Sustainable Development in South Asia. In I. Management Association (Ed.), Sustainable Development: Concepts, Methodologies, Tools, and Applications (pp. 1051-1087). Hershey, PA: IGI Global. doi:10.4018/978-1-5225-3817-2.ch047

Rizvi, S. M., & Dearden, A. (2018). KHETI: ICT Solution for Agriculture Extension and Its Replication in Open and Distance Learning. In U. Pandey & V. Indrakanti (Eds.), *Open and Distance Learning Initiatives for Sustainable Development* (pp. 163–174). Hershey, PA: IGI Global. doi:10.4018/978-1-5225-2621-6.ch008

Roşu, L., & Macarov, L. I. (2017). Management of Drought and Floods in the Dobrogea Region. In I. Management Association (Ed.), Agri-Food Supply Chain Management: Breakthroughs in Research and Practice (pp. 372-403). Hershey, PA: IGI Global. doi:10.4018/978-1-5225-1629-3.ch016

Rouzbehani, K., & Rouzbehani, S. (2018). Mapping Women's World: GIS and the Case of Breast Cancer in the US. *International Journal of Public Health Management and Ethics*, *3*(1), 14–25. doi:10.4018/IJPHME.2018010102

Roy, D. (2018). Success Factors of Adoption of Mobile Applications in Rural India: Effect of Service Characteristics on Conceptual Model. In M. Khosrow-Pour, D.B.A. (Ed.), Green Computing Strategies for Competitive Advantage and Business Sustainability (pp. 211-238). Hershey, PA: IGI Global. doi:10.4018/978-1-5225-5017-4.ch010

Sajeva, M., Lemon, M., & Sahota, P. S. (2017). Governance for Food Security: A Framework for Social Learning and Scenario Building. *International Journal of Food and Beverage Manufacturing and Business Models*, *2*(2), 67–84. doi:10.4018/IJFBMBM.2017070104

Sallam, G. A., Youssef, T., Embaby, M. E., & Shaltot, F. (2011). Using Geographic Information System to Infollow the Fertilizers Pollution Migration. In I. Management Association (Ed.), Green Technologies: Concepts, Methodologies, Tools and Applications (pp. 564-586). Hershey, PA: IGI Global. doi:10.4018/978-1-60960-472-1.ch312

Sambhanthan, A., & Potdar, V. (2017). A Study of the Parameters Impacting Sustainability in Information Technology Organizations. *International Journal of Knowledge-Based Organizations*, *7*(3), 27–39. doi:10.4018/IJKBO.2017070103

Sanga, C., Kalungwizi, V. J., & Msuya, C. P. (2017). Bridging Gender Gaps in Provision of Agricultural Extension Service Using ICT: Experiences from Sokoine University of Agriculture (SUA) Farmer Voice Radio (FVR) Project in Tanzania. In I. Management Association (Ed.), Discrimination and Diversity: Concepts, Methodologies, Tools, and Applications (pp. 682-697). Hershey, PA: IGI Global. doi:10.4018/978-1-5225-1933-1.ch031

Santra, A., & Mitra, D. (2017). Role of Remote Sensing in Potential Fishing Zone Forecast. In A. Santra & S. Mitra (Eds.), *Remote Sensing Techniques and GIS Applications in Earth and Environmental Studies* (pp. 243–257). Hershey, PA: IGI Global. doi:10.4018/978-1-5225-1814-3.ch012

Schaap, D. (2017). SeaDataNet: Towards a Pan-European Infrastructure for Marine and Ocean Data Management. In P. Diviacco, A. Leadbetter, & H. Glaves (Eds.), *Oceanographic and Marine Cross-Domain Data Management for Sustainable Development* (pp. 155–177). Hershey, PA: IGI Global. doi:10.4018/978-1-5225-0700-0.ch007

Seckin-Celik, T. (2017). Sustainability Reporting and Sustainability in the Turkish Business Context. In U. Akkucuk (Ed.), *Ethics and Sustainability in Global Supply Chain Management* (pp. 115–132). Hershey, PA: IGI Global. doi:10.4018/978-1-5225-2036-8.ch006

Segbefia, A. Y., Barnes, V. R., Akpalu, L. A., & Mensah, M. (2018). Environmental Location Assessment for Seaweed Cultivation in Ghana: A Spatial Multi-Criteria Approach. *International Journal of Applied Geospatial Research*, *9*(1), 51–64. doi:10.4018/IJAGR.2018010104

Sen, Y. (2018). How to Manage Sustainability: A Framework for Corporate Sustainability Tools. In I. Management Association (Ed.), Sustainable Development: Concepts, Methodologies, Tools, and Applications (pp. 568-589). Hershey, PA: IGI Global. doi:10.4018/978-1-5225-3817-2.ch026

Shamshiry, E., Abdulai, A. M., Mokhtar, M. B., & Komoo, I. (2015). Regional Landfill Site Selection with GIS and Analytical Hierarchy Process Techniques: A Case Study of Langkawi Island, Malaysia. In P. Thomas, M. Srihari, & S. Kaur (Eds.), *Handbook of Research on Cultural and Economic Impacts of the Information Society* (pp. 248–282). Hershey, PA: IGI Global. doi:10.4018/978-1-4666-8598-7.ch011

Sharma, Y. K., Mangla, S. K., Patil, P. P., & Uniyal, S. (2018). Analyzing Sustainable Food Supply Chain Management Challenges in India. In M. Ram & J. Davim (Eds.), *Soft Computing Techniques and Applications in Mechanical Engineering* (pp. 162–180). Hershey, PA: IGI Global. doi:10.4018/978-1-5225-3035-0.ch008

Silvestrelli, P. (2018). The Impact of Events: To Which Extent Are They Sustainable for Tourist Destinations? Some Evidences From Expo Milano 2015. In M. Risso & S. Testarmata (Eds.), *Value Sharing for Sustainable and Inclusive Development* (pp. 185–204). Hershey, PA: IGI Global. doi:10.4018/978-1-5225-3147-0.ch008

Silvius, G. (2018). Sustainability Evaluation of IT/IS Projects. In I. Management Association (Ed.), Sustainable Development: Concepts, Methodologies, Tools, and Applications (pp. 26-40). Hershey, PA: IGI Global. doi:10.4018/978-1-5225-3817-2.ch002

Singh, R., Srivastava, P., Singh, P., Upadhyay, S., & Raghubanshi, A. S. (2017). Human Overpopulation and Food Security: Challenges for the Agriculture Sustainability. In R. Singh, A. Singh, & V. Srivastava (Eds.), *Environmental Issues Surrounding Human Overpopulation* (pp. 12–39). Hershey, PA: IGI Global. doi:10.4018/978-1-5225-1683-5.ch002

Singh, R., Srivastava, P., Singh, P., Upadhyay, S., & Raghubanshi, A. S. (2017). Human Overpopulation and Food Security: Challenges for the Agriculture Sustainability. In R. Singh, A. Singh, & V. Srivastava (Eds.), *Environmental Issues Surrounding Human Overpopulation* (pp. 12–39). Hershey, PA: IGI Global. doi:10.4018/978-1-5225-1683-5.ch002

Srivastava, N. (2017). Climate Change Mitigation: Collective Efforts and Responsibly. In I. Management Association (Ed.), Natural Resources Management: Concepts, Methodologies, Tools, and Applications (pp. 64-76). Hershey, PA: IGI Global. doi:10.4018/978-1-5225-0803-8.ch003

Stanganelli, M., & Gerundo, C. (2017). Understanding the Role of Urban Morphology and Green Areas Configuration During Heat Waves. *International Journal of Agricultural and Environmental Information Systems*, *8*(2), 50–64. doi:10.4018/IJAEIS.2017040104

Stewart, M. K., Hagood, D., & Ching, C. C. (2017). Virtual Games and Real-World Communities: Environments that Constrain and Enable Physical Activity in Games for Health. *International Journal of Game-Based Learning*, *7*(1), 1–19. doi:10.4018/IJGBL.2017010101

Stone, R. J. (2017). Modelling the Frequency of Tropical Cyclones in the Lower Caribbean Region. In W. Ganpat & W. Isaac (Eds.), *Environmental Sustainability and Climate Change Adaptation Strategies* (pp. 341–349). Hershey, PA: IGI Global. doi:10.4018/978-1-5225-1607-1.ch013

Syed, A., & Jabeen, U. A. (2018). Climate Change Impact on Agriculture and Food Security. In A. Eneanya (Ed.), *Handbook of Research on Environmental Policies for Emergency Management and Public Safety* (pp. 223–237). Hershey, PA: IGI Global. doi:10.4018/978-1-5225-3194-4.ch012

Tam, G. C. (2017). The Global View of Sustainability. In *Managerial Strategies and Green Solutions for Project Sustainability* (pp. 1–24). Hershey, PA: IGI Global. doi:10.4018/978-1-5225-2371-0.ch001

Tam, G. C. (2017). Understanding Project Sustainability. In *Managerial Strategies and Green Solutions for Project Sustainability* (pp. 110–139). Hershey, PA: IGI Global. doi:10.4018/978-1-5225-2371-0.ch005

Tam, G. C. (2017). Perspectives on Sustainability. In *Managerial Strategies and Green Solutions for Project Sustainability* (pp. 53–76). Hershey, PA: IGI Global. doi:10.4018/978-1-5225-2371-0.ch003

Tang, M., & Karunanithi, A. T. (2018). Visual Logic Maps (vLms). In *Advanced Concept Maps in STEM Education: Emerging Research and Opportunities* (pp. 108–149). Hershey, PA: IGI Global. doi:10.4018/978-1-5225-2184-6.ch005

Taşçıoğlu, M., & Yener, D. (2018). The Value and Scope of GIS in Marketing and Tourism Management. In S. Chaudhuri & N. Ray (Eds.), *GIS Applications in the Tourism and Hospitality Industry* (pp. 189–211). Hershey, PA: IGI Global. doi:10.4018/978-1-5225-5088-4.ch009

Tianming, G. (2018). Food Security and Rural Development on Emerging Markets of Northeast Asia: Cases of Chinese North and Russian Far East. In V. Erokhin (Ed.), *Establishing Food Security and Alternatives to International Trade in Emerging Economies* (pp. 155–176). Hershey, PA: IGI Global. doi:10.4018/978-1-5225-2733-6.ch008

Tiftikçigil, B. Y., Yaşgül, Y. S., & Güriş, B. (2017). Sustainability of Foreign Trade Deficit in Energy: The Case of Turkey. In N. Ray (Ed.), *Business Infrastructure for Sustainability in Developing Economies* (pp. 94–109). Hershey, PA: IGI Global. doi:10.4018/978-1-5225-2041-2.ch005

Tiwari, S., Vaish, B., & Singh, P. (2018). Population and Global Food Security: Issues Related to Climate Change. In I. Management Association (Ed.), Climate Change and Environmental Concerns: Breakthroughs in Research and Practice (pp. 41-64). Hershey, PA: IGI Global. doi:10.4018/978-1-5225-5487-5.ch003

Toujani, A., & Achour, H. (2018). A Data Mining Framework for Forest Fire Mapping. In I. Management Association (Ed.), Information Retrieval and Management: Concepts, Methodologies, Tools, and Applications (pp. 771-794). Hershey, PA: IGI Global. doi:10.4018/978-1-5225-5191-1.ch033

Tripathy, B., & K., S. B. (2017). Rough Fuzzy Set Theory and Neighbourhood Approximation Based Modelling for Spatial Epidemiology. In I. Management Association (Ed.), *Public Health and Welfare: Concepts, Methodologies, Tools, and Applications* (pp. 1257-1268). Hershey, PA: IGI Global. doi:10.4018/978-1-5225-1674-3.ch058

Trukhachev, A. (2017). New Approaches to Regional Branding through Green Production and Utilization of Existing Natural Advantages. In I. Management Association (Ed.), Advertising and Branding: Concepts, Methodologies, Tools, and Applications (pp. 1758-1778). Hershey, PA: IGI Global. doi:10.4018/978-1-5225-1793-1.ch081

Tsobanoglou, G. O., & Vlachopoulou, E. I. (2017). Social-Ecological Systems in Local Fisheries Communities. In G. Korres, E. Kourliouros, & M. Michailidis (Eds.), *Handbook of Research on Policies and Practices for Sustainable Economic Growth and Regional Development* (pp. 306–316). Hershey, PA: IGI Global. doi:10.4018/978-1-5225-2458-8.ch026

Tsou, M. (2010). Geographic Information Retrieval and Text Mining on Chinese Tourism Web Pages. *International Journal of Information Technology and Web Engineering*, *5*(1), 56–75. doi:10.4018/jitwe.2010010104

Tuydes-Yaman, H., & Karatas, P. (2018). Evaluation of Walkability and Pedestrian Level of Service. In I. Management Association (Ed.), Intelligent Transportation and Planning: Breakthroughs in Research and Practice (pp. 264-291). Hershey, PA: IGI Global. doi:10.4018/978-1-5225-5210-9.ch012

Uddin, S., Chakravorty, S., Ray, A., & Sherpa, K. S. (2018). Optimal Location of Sub-Station Using Q-GIS and Multi-Criteria Decision Making Approach. *International Journal of Decision Support System Technology*, *10*(2), 65–79. doi:10.4018/IJDSST.2018040104

Uddin, S., Chakravorty, S., Sherpa, K. S., & Ray, A. (2018). Power Distribution System Planning Using Q-GIS. *International Journal of Energy Optimization and Engineering*, *7*(2), 61–75. doi:10.4018/IJEOE.2018040103

Uzun, F. V. (2018). Natural Resources Management. In A. Eneanya (Ed.), *Handbook of Research on Environmental Policies for Emergency Management and Public Safety* (pp. 1–21). Hershey, PA: IGI Global. doi:10.4018/978-1-5225-3194-4.ch001

V, M., Agrawal, R., Sharma, V., & T.N., K. (2018). Supply Chain Social Sustainability and Manufacturing. In I. Management Association (Ed.), *Technology Adoption and Social Issues: Concepts, Methodologies, Tools, and Applications* (pp. 226-252). Hershey, PA: IGI Global. doi:10.4018/978-1-5225-5201-7.ch011

van der Vliet-Bakker, J. M. (2017). Environmentally Forced Migration and Human Rights. In C. Akrivopoulou (Ed.), *Defending Human Rights and Democracy in the Era of Globalization* (pp. 146–180). Hershey, PA: IGI Global. doi:10.4018/978-1-5225-0723-9.ch007

Vaskov, A. G., Lin, Z. Y., Tyagunov, M. G., Shestopalova, T. A., & Deryugina, G. V. (2018). Design of Renewable Sources GIS for ASEAN Countries. In V. Kharchenko & P. Vasant (Eds.), *Handbook of Research on Renewable Energy and Electric Resources for Sustainable Rural Development* (pp. 1–25). Hershey, PA: IGI Global. doi:10.4018/978-1-5225-3867-7.ch001

Vázquez, D. G., & Gil, M. T. (2017). Sustainability in Smart Cities: The Case of Vitoria-Gasteiz (Spain) – A Commitment to a New Urban Paradigm. In L. Carvalho (Ed.), *Handbook of Research on Entrepreneurial Development and Innovation Within Smart Cities* (pp. 248–268). Hershey, PA: IGI Global. doi:10.4018/978-1-5225-1978-2.ch012

Wahab, I. N., & Soonthodu, S. (2018). Geographical Information System in Eco-Tourism. In S. Chaudhuri & N. Ray (Eds.), *GIS Applications in the Tourism and Hospitality Industry* (pp. 61–75). Hershey, PA: IGI Global. doi:10.4018/978-1-5225-5088-4.ch003

Weiss-Randall, D. (2018). Cultivating Environmental Justice. In *Utilizing Innovative Technologies to Address the Public Health Impact of Climate Change: Emerging Research and Opportunities* (pp. 110–143). Hershey, PA: IGI Global. doi:10.4018/978-1-5225-3414-3.ch004

Weiss-Randall, D. (2018). Cultivating Resilience. In *Utilizing Innovative Technologies to Address the Public Health Impact of Climate Change: Emerging Research and Opportunities* (pp. 204–235). Hershey, PA: IGI Global. doi:10.4018/978-1-5225-3414-3.ch007

Weiss-Randall, D. (2018). Climate Change Solutions: Where Do We Go From Here? In *Utilizing Innovative Technologies to Address the Public Health Impact of Climate Change: Emerging Research and Opportunities* (pp. 236–268). Hershey, PA: IGI Global. doi:10.4018/978-1-5225-3414-3.ch008

Whyte, K. P., List, M., Stone, J. V., Grooms, D., Gasteyer, S., Thompson, P. B., . . . Bouri, H. (2018). Uberveillance, Standards, and Anticipation: A Case Study on Nanobiosensors in U.S. Cattle. In I. Management Association (Ed.), Biomedical Engineering: Concepts, Methodologies, Tools, and Applications (pp. 577-596). Hershey, PA: IGI Global. doi:10.4018/978-1-5225-3158-6.ch025

Wulff, E. (2017). Data and Operational Oceanography: A Review in Support of Responsible Fisheries and Aquaculture. In P. Diviacco, A. Leadbetter, & H. Glaves (Eds.), *Oceanographic and Marine Cross-Domain Data Management for Sustainable Development* (pp. 303–324). Hershey, PA: IGI Global. doi:10.4018/978-1-5225-0700-0.ch013

Yener, D. (2017). Geographic Information Systems and Its Applications in Marketing Literature. In S. Faiz & K. Mahmoudi (Eds.), *Handbook of Research on Geographic Information Systems Applications and Advancements* (pp. 158–172). Hershey, PA: IGI Global. doi:10.4018/978-1-5225-0937-0.ch006

About the Contributors

José G. Vargas-Hernández, M.B.A., Ph.D., Member of the National System of Researchers of Mexico and a research professor at University Center for Economic and Managerial Sciences, University of Guadalajara. Professor Vargas-Hernández has a Ph. D. in Public Administration and a Ph.D. in Organizational Economics. He has undertaken studies in Organisational Behaviour and has a Master of Business Administration, published four books and more than 200 papers in international journals and reviews (some translated to English, French, German, Portuguese, Farsi, Chinese, etc.) and more than 300 essays in national journals and reviews. He has obtained several international Awards and recognition.

Justyna Zdunek-Wielgołaska is Assist. Prof. at Department of Polish Architecture, History of City Planning Lab at WUT. A graduate from the WUT, Faculty of Architecture in 2002, Ph.D. Diploma (with honors) in 2010. Awarded scholarships and research grants including Dean's grants and Research Project for doctoral dissertation granted by the Ministry of Science and Higher Education (2006). Beneficiary of Fulbright scholarship (visiting scholar at UC, Berkeley, USA, 2004-2005); beneficiary of the Institut Für Stadt- und Regionalplanung Technische Universität scholarship, Berlin, Germany and a laureate of the Start for Young Academics programme of the Foundation for Polish Science (2007). A graduate from the Post-graduate Estate Appraisal Studies (2003). Research experience (as leader): 1. Future of the large city suburbs. Smart suburbs – analysis for the Warsaw agglomeration (2017) 2. N-S artery and its role in functional and spatial shaping of interwar Warsaw. Idea and completion (2012) 3. Legal aspects of Warsaw public spaces shaping; premise and completion (2011) 4. Modernization of public spaces in the sense of city revitalization. Comparative study of the large American and Polish cities (2005).

* * *

Juan Esparza-López is a Research Student at University Center for Economic and Managerial Sciences, University of Guadalajara. Student of master in Business and studies in economy. Line of investigation is sustainable development.

Magdalena Grochulska-Salak is an architect, scientist and academic lecturer at the Faculty of Architecture Warsaw University of Technology also advisor for commercial investments, expert and project manager in an international development company. Scientific issues concern sustainable development in Eco City and Smart City, eco-production, urban farms in city, designing multi-family housing, designing and management of pro-environmental buildings in the city, buildings integrated with greenery.

Paula Guerrero Rodríguez has a Degree in Tourism University of Guadalajara. Master's Degree in Higher Education at the University Center of La Ciénega, University of Guadalajara. Doctorate in School and Educational Administration from the Mexican Institute of Pedagogical Studies. Guadalajara Jalisco. Research Coordinator at the University Center for Economic and Administrative Sciences (CUCEA). Professor of the Degrees of Human Resources, International Business, Administration and Marketing in the University Center of Economic and Administrative Sciences (CUCEA). Professor of the Degree in Tourism at the University Center for Economic and Administrative Sciences (CUCEA). Visiting Professor at the National University of Jujuy, Argentina to teach a postgraduate course "Specialization in Higher Education" called "Organization and Educational Administration". Presenter of diverse investigations at countries such as: Peru, Argentina, Cuba, Spain and Mexico. Publications: 14 refereed articles 1 chapter of a book.

Krystyna Guranowska-Gruszecka, Ph. D. Eng., of Warsaw Technical University, Department of Architecture, Chair of Urban Planning and Rural Landscape. Between 2006 and 2015 the head of doctoral studies programme of the Architecture Department. An active lecturer and tutor in Wyższa Szkoła Przedsiębiorczości in Lublin and Społeczna Akademia Nauk SAN in Warsaw. An experienced urban planner with several hundred local development plans, development condition studies, directions for development, and other urban projects in her portfolio. Author of several books, including a monography "The Centre of Warsaw in the 20th Century".

Mariko Ikeda received Doctor of Science from the faculty of Life and Environmental Sciences in University of Tsukuba. Through her international academic experiences in Germany, she is an assistant professor in the Faculty of Art and Design, University of Tsukuba. Her academic interest lies in the urban transformation with various forms, such as gentrification, urban regeneration and sports related Mega-event. More recently, she has published article about Night-time Economy, which emerges in pre-Olympics of Tokyo, Japan.

Erhan İşcan received his Ph.D. degree in Economics from the Cukurova University. Dr. Erhan İşcan is an Assistant Professor at Cukurova University. His interests are focused on Economics Theory, Digital Economy, Space Economy and Bioeconomy.

Baseem Khan received his Bachelor of Engineering degree in Electrical Engineering from the Rajiv Gandhi Technological University, Bhopal, India in 2008. He received his Master of Technology and Doctor of Philosophy degrees in Electrical Engineering from the Maulana Azad National Institute of Technology, Bhopal, India, in 2010 and 2014, respectively. Currently, he is working as an Assistant Professor in the School of Electrical and Computer Engineering, Hawassa University Institute of Technology, Hawassa University, Hawassa, Ethiopia. His research interest includes power system restructuring, power system planning, smart grid technologies, meta-heuristic optimisation techniques, reliability analysis of renewable energy system, power quality analysis and renewable energy integration.

Kai Liao is a student in the graduate school of Marine Science and Technology, Tokyo University of Marine Science and Technology.

Shimon Mizutani is a Doctoral Student in the Graduate School of Marine Science and Technology, Tokyo University of Marine Science and Technology.

Rogelio Rivera-Fernández is an Electrical Engineer and Law Degree. Master of Administration and Education. Doctor in Administrative Sciences from the National Polytechnic Institute and professor and researcher at the University of Guadalajara, Mexico. Dr. Rivera is Senior Research Professor "C" with Desirable Profile Member of the SNI National System, with level I. Accredited as visiting professor at the Doctorate in Administrative Sciences of the ESCA-IPN, at México city. Evaluation member of undergraduate and postgraduate Academic Programs in the subgroup of Economic and Administrative Sciences.

Tsuyoshi Sasaki is a Professor, Graduate school of Marine Science and Technology, Tokyo University of Marine Science and Technology.

Feng Wu is a Professor in the Architecture College of Xi'an University of Architecture and Technology.

Index

A

ANSYS 65, 69, 73-74, 76-79, 81, 84

B

Bio economy 2-3, 6, 9-12
Bio economy strategies 10
Biobased Products 26, 33
Bio-Based Products 3, 42
Biocombustibles 115
Biodiesel 57, 89-99, 101-102, 106-110, 115
Bio-Ecology 4, 8-9, 14, 20
Bioeconomic Valuation 178
Bioeconomy Vision 31
Bio-Refinery 20
Bio-Resources 3, 6-8, 20
Biotechnology 20, 25-30, 32-37, 176-177
Bio-Technology 4-8, 10-11, 13-14, 20
Blue Economy 42

C

Center of Warsaw 117, 119, 122, 124, 127, 130, 134-135, 138, 142, 148
Circular Economy 42
Cities 22-23, 46-48, 50-51, 55, 57-62, 118, 120-121, 136, 141, 144-145, 149, 151, 153, 165, 171-172, 178-179, 181, 197, 200, 203, 206, 208, 212
Company 97-99, 115, 127, 162, 174, 214
COMSOL 65, 69-70, 73-74, 76-79, 81, 84
Construction of Beautiful Village 181, 195

D

DWG File 189, 195

E

Eco City 47, 59, 64
Eco-City 47, 50, 59, 64, 196, 199-200, 216
Economics 31, 35, 42, 46-47, 124, 133, 241
Economy 1-3, 6, 9-13, 22-24, 27-28, 31-34, 37, 42, 47-48, 57-58, 90, 100, 102-103, 106, 109-110, 133, 138, 152-153, 193, 202, 217
Eco-production 44, 50, 60-61
Entrepreneurship 89-90, 101-104, 106, 109, 115, 163
Environmental policies 171, 200-202, 207-208

F

Food Security 11, 22, 31-32, 42, 59, 92, 175-176
Formal Institutions 89-92, 97, 99, 101, 104, 107, 109-111, 115

G

geographical data 184, 189
Green City 64
Green Economy 42
Green Urban Network 148
Green Ventilating Wedges 146, 148

Greenery 43-45, 51-52, 54-55, 57, 59, 64, 117-118, 120-124, 126, 129-131, 133-136, 138-146, 203
Greening Strategy 121, 149
Guanzhong 180-184, 187, 192-193, 195
Guanzhong region 180, 184

H

Health 2, 8, 27, 29-30, 32, 34, 49, 118-119, 121, 133-134, 141, 146, 149, 155, 158, 171, 175-177, 225, 236, 240
Heat Island 196-197, 199, 201, 208-212, 217

I

Informal Institutions 92, 99-101, 104, 115
Insecurity 151, 171, 173, 179
Institution 13, 35, 99, 115
Institutions 12-13, 25, 35-36, 89-92, 96-101, 104-105, 107-111, 115, 131, 146, 155, 166, 168, 173

K

Knowledge-Based Bioeconomy 27-28, 31-33, 42

M

Market 5, 13-14, 33-34, 36, 89-92, 95-98, 100, 103, 107-111, 133, 141
México 57, 89, 91-97, 99, 101, 105-110, 115, 151, 155, 160, 166, 170-171, 174

N

Natura 2000 149

O

OECD 2-3, 10, 24-26, 29-36, 38, 90, 102, 175, 201
Olympic Games 199, 202-203, 205-206, 208, 216
Oxygen 55-56, 61

P

Piezoelectric material 65-67, 69, 84-85
Policy mix 2, 14

Q

Quintuple Helix 152, 155, 174, 179

R

Renewable Biomass 26, 28-29, 42
Rooftop Greening 212, 214

S

Smart City 44, 47-48, 50-51, 59, 61, 64
Society Exclusion 179
Stress 67, 71, 79, 81, 151, 179
Sustainable City Development 14, 43
Sustainable Development 2, 14, 22, 26-27, 38, 43-46, 48-52, 54-55, 57-61, 64, 122, 152-153, 197, 199, 205, 207, 230-231, 237, 240

T

Tokyo 52-53, 196-197, 199-200, 202-212, 216-217
Traffic Generators 121, 149

U

UAV 180, 185-189, 192-193
Urban Agriculture 44, 54, 64
Urban Climate 199
Urban Era 196-197, 199, 216
Urban Farming 43
Urban Green Areas 158, 165, 179
Urbanism 179

W

Wall Greening 196, 209-211, 213-214, 217
Warsaw 49, 55, 57, 117-118, 120-124, 127, 130-135, 137-143, 145-149

www.ingramcontent.com/pod-product-compliance
Lightning Source LLC
LaVergne TN
LVHW081037190926
848465LV00005B/121

* 9 7 8 1 5 2 2 5 7 9 5 8 8 *